How Do I Do That In My Tesla?

The Quickest Ways to Do the Things You Want to Do, Right Now!

Scott Kelby
& Terry White

How Do I Do That In My Tesla?

The *How Do I Do That In My Tesla* Book Team

MANAGING EDITOR
Kim Doty

COPY EDITOR
Cindy Snyder

ART DIRECTOR
Jessica Maldonado

PHOTOGRAPHY
Scott Kelby
Terry White

PUBLISHED BY
Rocky Nook
Rocky Nook, Inc.
1010 B Street, Suite 350
San Rafael, CA 94901
info@rockynook.com
(415) 747-8756

Composed in Univers and Avenir (Linotype) by KelbyOne.

ISBN: 979-8-88814-376-6
10 9 8 7 6 5 4 3 2 1

Printed in Dubai
This book is printed on acid-free paper.

Distributed in the UK and Europe by Publishers Group UK
Distributed in the U.S. and all other territories
by Publishers Group West

www.kelbyone.com
www.rockynook.com

This book is dedicated to our dear friend and colleague Erik Kuna, who was the first guy we knew with a Tesla. He made us believers, then owners, and we've never looked back. Thanks, Erik, for rekindling our love of driving and making it all fun again.

Scott's Acknowledgments

Although only two names appear on the spine of this book, it takes a team of dedicated and talented people to pull a project like this together. I'm not only delighted to be working with them, but I also get the honor and privilege of thanking them here.

To my amazing wife Kalebra: You continue to reinforce what everybody always tells me—I'm the luckiest guy in the world.

To my son Jordan: If there's a dad more proud of his son than I am, I've yet to meet him. You are just a wall of awesome! So proud of the fine young man you've become. #rolltide!

To my beautiful daughter Kira: You are a little clone of your mom (except for your love of horror movies), and that's the best compliment I could ever give you.

To my big brother Jeff: Your boundless generosity, kindness, positive attitude, and humility have been an inspiration to me my entire life, and I'm just so honored to be your brother.

To my editor Kim Doty: I feel incredibly fortunate to have you as my editor on these books. In fact, I can't imagine doing them without you. You truly are a joy to work with.

To my book designer Jessica Maldonado: I love the way you design, and all the clever little things you add to everything you do. Our book team struck gold when we found you!

To my dear friend and business partner Jean A. Kendra: Thanks for putting up with me all these years, and for your support for all my crazy ideas. It really means a lot.

To Erik "The Real Rocket Man" Kuna: For taking the weight of the world on your back, so it didn't crush mine, and for working so hard to make sure we do the right things, the right way.

To Ted Waitt, my awesome "Editor for life" at Rocky Nook: Where you go, I will follow.

To my publisher Scott Cowlin: I'm so delighted I still get to work with you, and thankful for your open mind and vision.

To my mentors John Graden, Jack Lee, Dave Gales, Judy Farmer, and Douglas Poole: Thank you for your wisdom and whip-cracking—they have helped me immeasurably.

Thanks to these folks who had nothing to do with this book, but so much to do with my life: Paul Kober, Jeff Revell, Juan Alfonso, Moose Peterson, Larry Grace, Rob Foldy, Dave Clayton, Victoria Pavlov, Dave Williams, Kelly Jones, Christina Sauer, Kleber Stephenson, Larry Becker, Peter Treadway, Roberto Pisconti, Fernando Santos, Chris Main, Marvin Derizen, Maxx Hammond, Brad Moore, Joe McNally, Jason Stevens, Annie Cahill, Rick Sammon, Mimo Meidany, Tayloe Harding, Dave Black, John Couch, Greg Rostami, Frank Doorhof, Peter Hurley, Kathy Porupski, and Vanelli.

Most importantly, I want to thank God, and His Son Jesus Christ, for leading me to the woman of my dreams, for blessing us with such amazing children, for allowing me to make a living doing something I truly love, for always being there when I need Him, for blessing me with a wonderful, fulfilling, and happy life, and such a warm, loving family to share it with.

Terry's Acknowledgments

There's a lot that goes into a book like this and although you just see the names of the authors on the cover, I couldn't have done it without the people that inspire me and push me to do better on a daily basis.

To my beautiful wife Victoria: You continue to make me want to be a better person and be the best that I can be. I love you!

To my daughters Ayoola and Sala: Who have collectively read more books that I ever will. It feels good to write one that I think you would be proud of.

To my big sister Pam: You have always been there for me when I needed you. You are always smiling and always positive. Thank you for being you.

To my cousin Nita: Who always gets excited when I show her new tech. I love geeking out with you.

To my best friend Scott Kelby: I want to thank you for all that you have done through the years to show me that there are truly good people in this world. You're an inspiration and the greatest friend that anyone could have.

To my friend Sergio Rodriguez: I can't keep up with your EV collection. You inspire me by always pushing the envelope on driving all different brands of EVs and showing us that not only is owning and driving an EV more fun, but also that cross-country EV trips are a piece of cake. I've lost count of how many you've done.

To Shelly and Terry Travis: Thank you inviting me to my first EV group and not only making me feel welcome, but for all the work you do to educate the underserved communities on EV ownership.

To Dave Helmly: Dave, you were one of the first Tesla owners I knew and it's been a blast following your journey with not only your Teslas, but also your Tesla Solar Roof. I also love the story about how you surprised your wife with a new Tesla that she thought she was just going to look at, but was actually her new car. Well played, my friend!

To Larry Becker: You're one of my favorite people to talk tech with and although you don't drive a Tesla yet, you're always there to bounce ideas off of. Plus, every now and then, you see new tech that you know I'll be interested in before I see it and share it with me. Thanks Larry!

I would be remiss if I didn't thank all the people that keep me going every single day: Jack Beckman, Rufus Deuchler, Kurtis Wilkins, DeMon Lewis, Jason Levine, Greg Wilson, Christa Lee, Shatanese Reese, Steve Reese, Terrence Turner, Asher Turner, Alcia Jordan, Glyda Archer, Lise Blaise, Jordan Ellis, Elise Swopes, Katrina Torrijos, Teresa Au, Michael Fuguso, Sydney Lanham, Lo Pelusi, David McGuire, Mia Sasser, Rosalia Ley Asia Lewis, Selima Benjamin, Rudy Benjamin, Cynthia Smith, Penny Smith, Jerry Silverman, Queenie White, Tara Abtahi, Juanita Ruffin, Maurice, Chita Hunter, Aundre Larrow, Jerlyn Thomas, Kelsey Slay, Joe Smith, Russell Preston Brown, Chris Ayers, Samantha Shoushtari, Brandon Heiss, Gustavo Brunser, Iris Pérez, Jolisa Copeman, W. Taharqa Blue, Shana Dezelle, Renee Ferguson, Dave Clayton, Sloka Akula, Eric Cornish, and Shane Whatley.

About the Authors

Scott Kelby

Scott is President and CEO of KelbyOne, an online educational community for photographers. He is the host of *The Grid*, the influential live weekly talk show for photographers, and is founder of the annual Scott Kelby's Worldwide Photo Walk.™

Scott is an award-winning photographer, designer, and bestselling author of more than 100 books, covering everything from technology to being a dad. He is a long-time Tesla enthusiast, starting with his first Model S nearly 10 years ago, and now between his wife's and their daughter's first Tesla Model 3, they are now on their fourteenth Tesla.

Scott is a frequent speaker at conferences and trade shows around the world and is featured in a series of online learning courses at KelbyOne.com.

For more information on Scott, visit him at:
X: **@scottkelby**
Facebook: **facebook.com/skelby**
Instagram: **scottkelby**
Threads: **@scottkelby**
LinkedIn: **scottkelby**

Terry White

Terry White is a renowned technology expert, author, and award-winning photographer, with a deep passion for Tesla, electric vehicles, and innovation. With over 28 years of experience in the tech industry, including his influential role as a Principal Director at Adobe, Terry brings a unique blend of technical expertise and creative vision to the world of EVs.

As a Tesla owner and enthusiast, Terry has become a trusted voice in the EV community, sharing insightful reviews, practical tips, and Tesla how-tos through his popular YouTube channel. His tech blog delves into the transformative impact of Tesla and electric vehicles, offering readers a comprehensive guide to embracing the EV lifestyle and understanding the technology shaping our future.

A sought-after speaker at events like TEDx and Adobe MAX, Terry combines his extensive technical background with a forward-thinking perspective, inspiring audiences worldwide to explore the exciting possibilities of sustainable transportation and smart technology.

For more information on Terry, visit him at **terrywhite.com** or:
YouTube: **youtube.com/@TerryLeeWhite**
Instagram: **terryleewhite**
Facebook: **facebook.com/terrywhitefans**
X: **@terryleewhite**

Table of Contents

Chapter 1

How to Do the Essentials 1

Stuff You Gotta Know Up Front

Table of Contents

Table of Contents

Table of Contents

Table of Contents

Table of Contents

Table of Contents

Table of Contents

Table of Contents

Table of Contents

Table of Contents

Chapter 12

How to Deal with the Haters 277

Owning a Tesla Does Not Come without Its Challenges

If You Read This First, We Promise, It Will Help a Lot

(1) This book is geared toward Model 3 and Model Y owners. These two models are very similar (even though their body styles are different), but where there's a difference between the two, we point it out. We do include some topics, or show how a feature or control is different on the Model X or Model S, but just so you know up front, the book is really aimed at Model 3 and Y owners. While we're mentioning the Model X and S, it's possible you ordered yours with a yoke instead of a steering wheel, but in the book, we always refer to it as a steering wheel. It just makes things easier.

(2) The Model 3 had a major update in 2024 and the Model Y got one in 2025. They updated the front and back of both models, including the interiors, added new features, added a new screen in the back seat—both saw pretty big updates (inside and out). While the book is based on these newer models, we still include how to do stuff on older Model 3s and Ys.

(3) Don't read the book in order—it's not that kind of book. This is more like an "I'm stuck. I need to know how to do something in my Tesla right now" book. So, when you need to know how to do a particular thing right this very minute, you just pick it up, turn to the chapter where it would be (Climate, Service, Navigation, etc.), find the thing you need to do, and we tell you exactly how to do it—all on just one page and pretty succinctly ($5 word—bonus points!). If we did our job right, you should only be in this book for like a minute or two at a time—just long enough to learn that one important thing you need to know right now, and then you're back to lounging on your yacht (at least, that's how we imagine your life will roll after buying this book).

A Few More Important Things You'll Want to Know

(4) We wrote this book like we were talking to a friend. We talk throughout the book like we're sitting with a cup of coffee talking with one of our friends about how to get the most out of their Tesla. So, we leave out a lot of the nerdy techno-jargon and just tell it to you in plain English, like we would to a friend.

(5) It's possible some new features have been added since this book was written. Tesla's frequent over-the-air updates, which often bring new features or new options, are one of the best things about owning a Tesla (and one of the worst things about being a Tesla author), so it's possible there are a few features we haven't covered, but that's because they snuck in while you (and we) were sleeping one night. But, don't worry, we've got you covered, as we created a webpage for the book where we're going to add new features as they're released. So, go to **kelbyone.com/books/teslabook** to check on any new feature releases.

(6) There's a bonus chapter at the end of the book. Once you own a Tesla, you'll meet a lot of people who are interested or curious about it, and while these folks are generally well-intentioned, they have undoubtedly heard on TikTok or Facebook a bunch of things about owning a Tesla that are anywhere from slightly off to just flat-out crazy wrong. These people do not seem shy about chatting you up about it in the parking lot of your local grocery store. Knowing that, we addressed the most common misconceptions in this bonus chapter, so when they say crazy things to you (like the battery has to be replaced every couple of years—yes, we've heard that one), you'll have the real facts at the ready.

Most Everything Happens on This One Screen

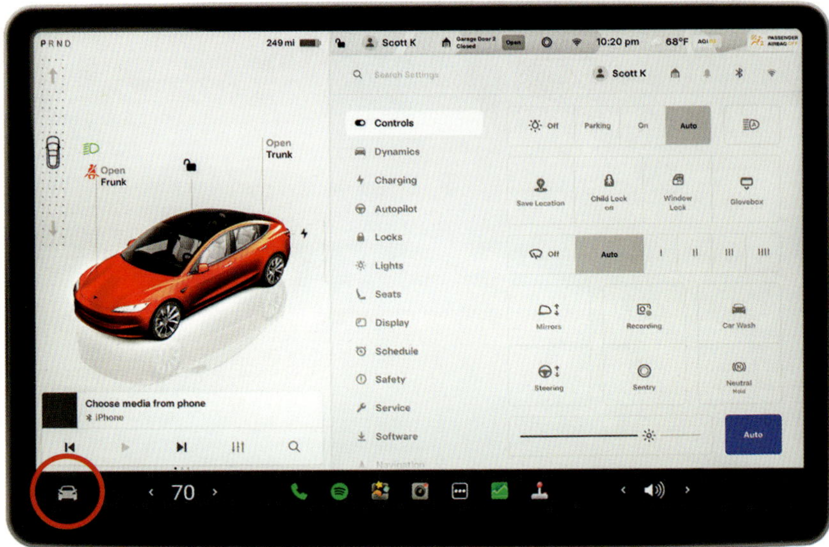

(7) There are very few physical buttons in your Tesla. That's because everything pretty much happens on your center touchscreen (seen above). Your Tesla operates a lot like your phone. There are a few buttons on your phone, but most of the time you do everything by tapping on your phone's screen, right? Well, think of your Tesla's center touchscreen as a bigger version of your phone's screen (more like an iPad) because the two work very much the same (your Tesla even has its own apps). For example, on your phone, most of your options and preferences are found in its Settings app. Your Tesla has something similar called "Controls." Most of the options and preferences for your car are found by tapping on the Controls icon, which lives in the bottom left of your center touchscreen (it's the car icon circled above in red). When you tap on that icon and its screen appears, you'll see buttons for the things you're most likely to use on a day-to-day basis (things like headlights, wipers, folding in your side mirrors, etc.). Along the left side of the Controls screen is a list of all the other things you might not need to adjust daily, but that let you customize your Tesla experience to your own taste. Essentially, the Controls icon is your gateway to setting up everything your way. Also, along the bottom of your touchscreen is an area called the "My Apps" bar, where you can make some of your favorite Tesla apps always visible and just one tap away (apps like Spotify, or your phone, games, FM radio—stuff like that). You'll learn how to add your favorites to that bar later (on page 37), but for now we thought it might be helpful to know what a big role your center touchscreen plays in your day-to-day driving experience (plus, you'll hear us talking about it a lot throughout the book).

There's Just a Little Bit More. Keep Reading

(8) From Scott: I do this thing that either delights my readers or makes them spontaneously burst into flames of anger, but it has been a tradition of mine for more than 25 years of writing books, so now it's a "thing." It's how I write the chapter intros. In a normal book, they would give you some insight into what's coming in the chapter. But, mine...um...they have little, if anything, to do with what's in the chapter ahead, as I've designed them to simply be a "mental break" between chapters, and these quirky, rambling intros have become a trademark of mine (in fact, if you look on Amazon, you'll find an entire book of nothing but my chapter intros. No, I am not making this up). Luckily, the rest of the book is pretty regular, but I had to warn you about these intro pages, just in case you're one of those serious types (AKA: Mr. Grumpy Pants). If that sounds like you, I'm begging you, please skip the chapter intros—they'll just get on your nerves.

Okay, that's pretty much it. You're prepped and you're now ready to dive into the rest of the book (and thanks for reading these four pages—I know you wanted to get right to it, so a big high-five for sticking with us here first). Terry and I hope you find this book really helpful in your Tesla journeys. Let's start her up!

MODEL3
SUNSHINE STATE

Chapter 1

How to Do the Essentials

Stuff You Gotta Know Up Front

Let's start this first chapter intro off with a quiz. Now, you might think that this is a bit early for a Tesla pop quiz, seeing as you haven't read any pages yet, but I have good news: this is not about your Tesla (which is a good concept to keep in mind as you move through subsequent chapter intros). This is more about reading comprehension, but don't worry, you'll be happy to learn that it's only a two-question quiz, and there is no wrong answer (unless you choose "B," which is the wrong answer). This is a timed test, and you are not allowed to use your calculator. Also, all answers should be submitted in writing two weeks prior to the test. Ready? You may begin. Q. Which of these statements is true: (A) I read page xxi in the introduction of this book and fully understand that these quirky chapter introductions have little, if anything, to do with what's actually in the chapter ahead. I acknowledge that they are simply designed to give readers, like me, a mental break between chapters, which is what makes this first chapter introduction so vexing as I haven't read any Tesla stuff at this point, but yet I'm receiving a mental break. But, I'm okay with all this. I'm chill. I don't anger easily and I wouldn't let a paragraph of incoherent ramblings at the beginning of a chapter affect my enjoyment of this chapter to any major extent. Or, (B) I find any minor interruption in my quest for Tesla proficiency, no matter how brief, highly annoying. I have little tolerance for things like chuckling, laughing, or giggling, nor do I have time for superfluous things like sunny days, rainbows, or puppies. The most important thing for me is the pursuit of Tesla knowledge, so I can go to online Tesla forums and poke fun at people who ask simple questions about their Tesla. My favorite thing is answering their questions using features that don't actually exist. For example, if they ask: "How do I open the front trunk?" I might answer: "Go to the center touchscreen, tap on Controls, then tap on HoodsaPoppin'" (which does not exist). Then, when they comment that they can't find the HoodsaPoppin' button, my colleagues and I publicly poke fun at them until that person deletes their account. See, *that* I find pleasurable. So, I choose "B." The answer was "A," though, wasn't it? I knew it. This test is rigged!

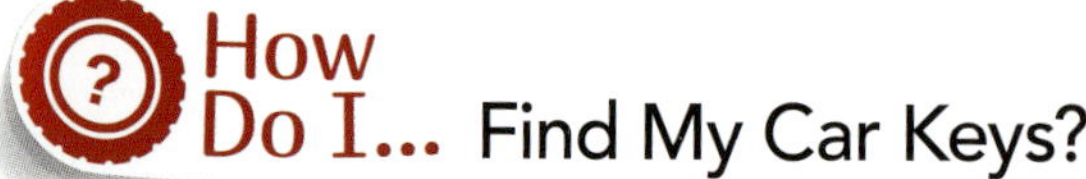

How Do I... Find My Car Keys?

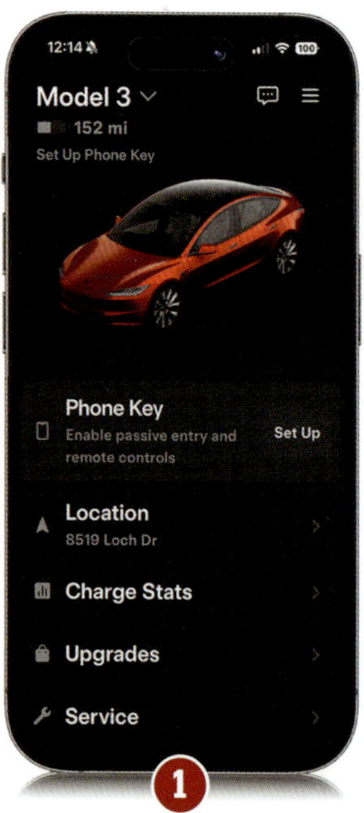

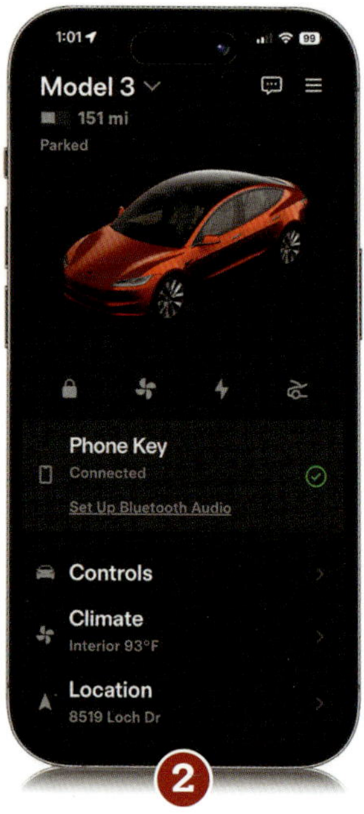

This might sound weird at first, but your phone is your car key and it's connected to your car using Bluetooth. That's right—as you move toward your car and then press the door handle, the doors automatically unlock so you can get in and drive. They set this up for you when you take delivery of your car at the Tesla Service Center, but if they didn't for some reason (or you bought a used Tesla from an individual, so there was no Tesla staff involved), you can set this up yourself by going up to the driver's side of your car and launching the Tesla app on your phone, where you'll see a list of options. First, tap on Phone Key. Right beneath that you'll see "Enable passive entry and remote controls," and to the right of that is a Set Up button (as seen above in #1). Tap that Set Up button and the app will start looking for your Tesla. (*Note:* You'll need to be physically close to your car when you do this.) When it finds it, you'll see "Phone Key Connected" (as seen above in #2). That's it—your phone now acts as your car key. But, what's nice is, you don't even have to take your phone out of your pocket (or purse) to open the doors or drive. It all works by Bluetooth, so once you get close to your car, it unlocks—just get in and start driving (there's nothing else to do key-wise). One more thing: Although your phone is your key now, your phone isn't connected to the audio to play music from your phone. If you want to do that, tap on Set Up Bluetooth Audio (right beneath Phone Key Connected). If you forgot to do that, no worries, you can add it later (see page 35).

How Do I... Get In If My Phone Won't Unlock My Tesla?

This can happen if your phone's battery is dead (it happened to me coming home from a long flight), or your phone's Bluetooth feature is turned off or just isn't working, or if something's just weird (rare, but weird stuff can happen with wireless stuff). This is why you get two key cards when you buy a Tesla. They are the size and thickness of a credit card (maybe a little thinner—more like a hotel room key) and you can use your key card to unlock your car by waving it right up against the center pillar between the front and back windows. You might see an oval with a camera inside it inside the pillar and it would seem like that's where you would wave your card, but that's actually not the spot (it won't work up there). Wave it about four or five inches lower—at about the center of the pillar—and place the key card right up against the pillar so it's touching it. That'll do the trick—you'll know it's unlocked because the side mirrors will unfold. If it doesn't unlock, it's most likely because you're either holding the card up too high on the pillar or you're not holding it close enough. Because the key card is the easiest way to get into your car (if your phone doesn't work for some reason), I highly recommend keeping one in your wallet or purse. (*Note:* There's another way to unlock your Tesla—if you don't have your key card—using the Tesla app, which Terry covers on page 169.) One more tip: turn the key card over (so you're not seeing the Tesla logo side of it) and you'll find a little diagram that shows you where to place the key card to unlock your car. You cannot imagine how many people have seen that diagram and had no idea that's what it was for. You can't make this stuff up, folks (well, you could, but you usually don't need to).

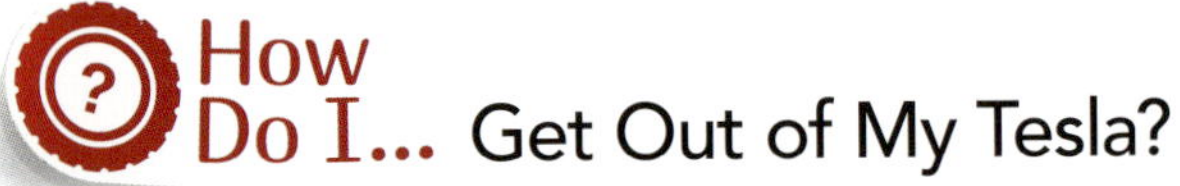

How Do I... Get Out of My Tesla?

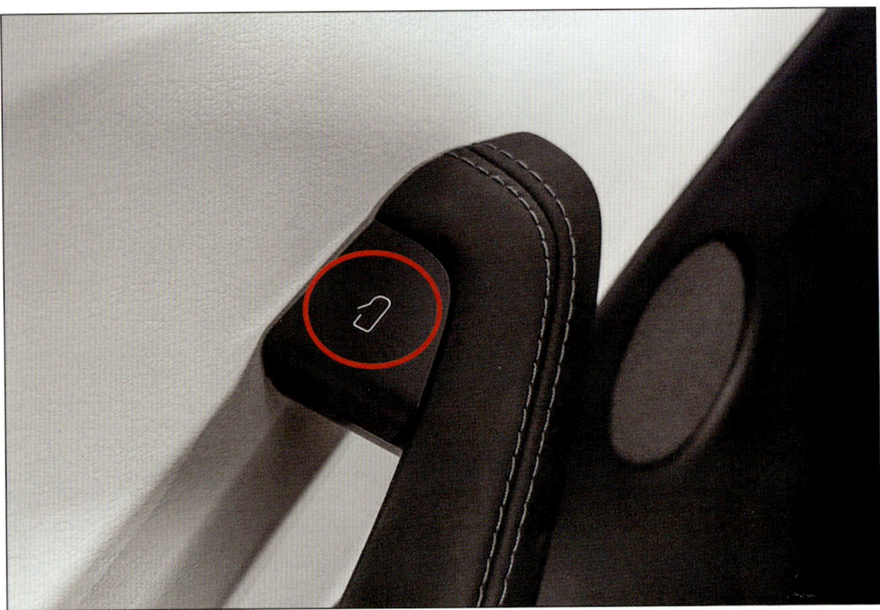

There isn't a door handle like we're used to using. Instead, there's a single open door button that pops the door open a bit for you, and then you can swing it open like any other car door. It's the button on the inside of the door handle that has a little icon that looks like an overhead view of your car with an open door. Press that button once, the door pops open a bit, and now you can open it the rest of the way.

How Do I... Get Out If the Door Button Doesn't Work?

If you press the open door button and for some reason it doesn't work (that's rare, but it could possibly happen, like in the unlikely case that your car has no power), there's a manual door release right in front of the window buttons. Reach your fingers beneath the area in front of those buttons, pull that up, and it manually opens the door. You'll get a warning on your center touchscreen that you manually opened the door, which warns you for two reasons: (1) a passenger could've accidentally opened a door, so it's letting you know, and (2) it's not the recommended way to open the door (you should use the door button). (*Note:* See page 253 for another manual way to do this.)

How Do I... Adjust the Seats?

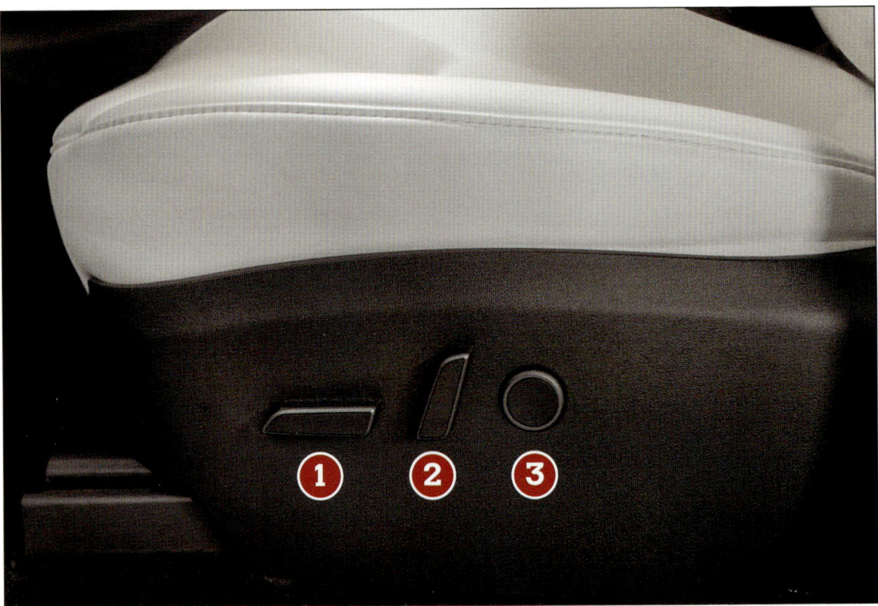

All Teslas have electric seats and the controls appear on the bottom-left side of the driver's seat and the bottom-right side of the passenger's seat—just reach down there and you'll feel them. The first control (#1 above) is for the base of the seat and is shaped just like it, which makes it pretty easy to get your seat just the way you want it. If you want your seat lower, press down on this button. If you want it higher, pull it upward. If you want your seat closer to the steering wheel, push the button forward. If you want it farther away, push it backward. Pretty simple. The next control, in the middle (#2 above), controls the back of your seat and its angle. To lean your seat backward, press it backward. To have it more straight, press it forward toward the front of the car. The last control, the round one (#3 above), is for adjusting the lumbar support. As you press the top, bottom, left, or right side of this button, you'll feel the lumbar support in your back, so you get instant feedback. Of course, you can save these settings so each time you unlock the car, it automatically puts the seat in your favorite position. This is handy if multiple people drive the car. For example, my wife and I have separate profiles, so when she's going to drive, and unlocks the car with her phone, it automatically switches the seats, mirrors, and entertainment features—all that stuff—to her personal preferences. For more on how to save these to your own driver's profile, head over to page 10.

How Do I... Adjust the Front Passenger's Seat?

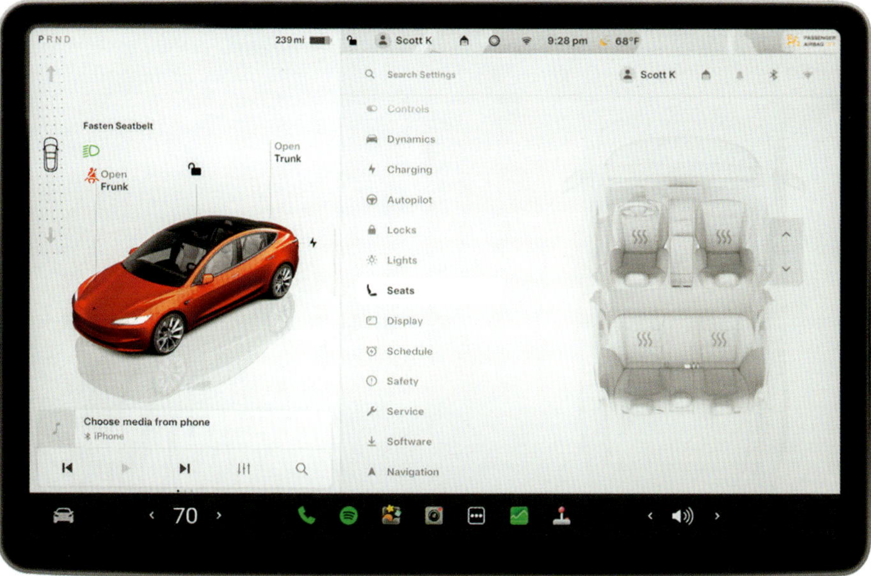

This is great if you're an Uber driver and want to move the front passenger's seat all the way forward, so your back seat client has more room (okay, sorry—I couldn't help myself). To adjust the position of the front passenger's seat, just tap on the Controls icon (the car in the bottom left of your center touchscreen), and then tap on Seats. To the right of the seat, you'll see forward and backward arrow buttons you can tap to adjust how far up or far back you want the seat.

How Do I... Adjust My Mirrors?

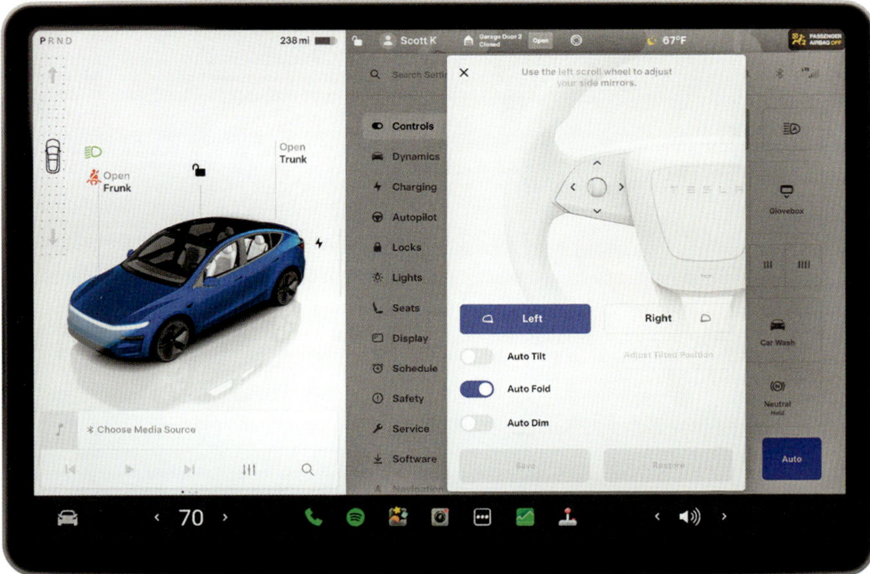

On your center touchscreen, tap on the Controls icon (the car in the bottom left), then tap on the Mirrors button, and then tap to choose either the left or right mirror. Once you've done that, you'll actually use the left scroll button on your steering wheel to move the mirrors up/down or left/right. Roll the scroll button up/down to move the mirror up or down, and drag the button (well, move it with your finger/thumb) left/right to move the mirror to the left or right. Once you've got your mirrors how you like them, save those settings to your driver's profile, so that anytime you get into your car, it will automatically make sure your mirrors are set to where you want them. To learn how to save these to your driver's profile, see page 10.

How Do I... Adjust the Position of My Steering Wheel?

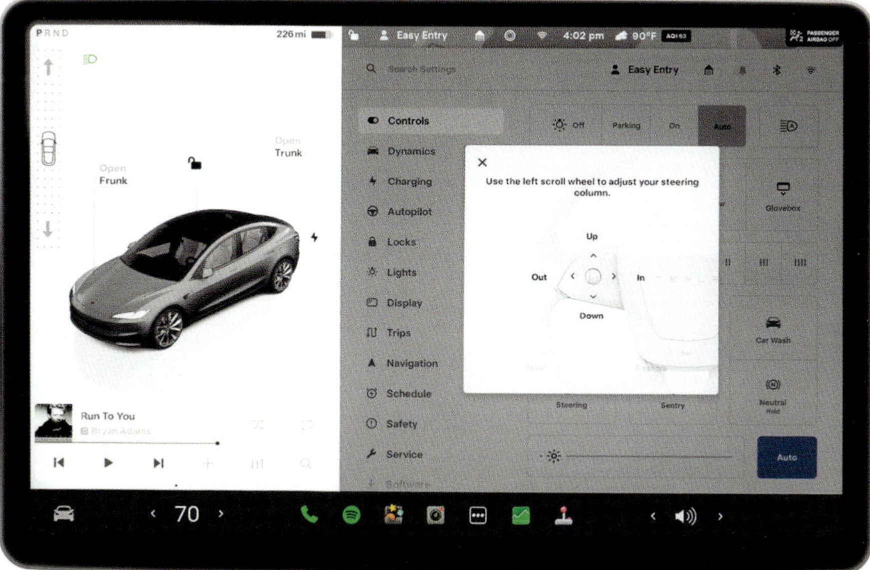

Go to your center touchscreen, tap on the Controls icon (the car in the bottom left), and then tap on the Steering button (it has a steering wheel on it). Now, to adjust its position, use the left scroll button on the steering wheel itself. Roll the scroll button up/down to move the steering wheel up or down, and move the button left/right to adjust how close the steering wheel is to you.

TIP: EXPANDING YOUR SUN VISOR

You can extend the length of your sun visor (in case it's not blocking enough of the sun) by grabbing the end of the visor and pulling outward. The visor extends by sliding the entire visor out a few inches.

How Do I... Save My Seat Position, Mirrors, Steering Wheel Position, and Other Preferences?

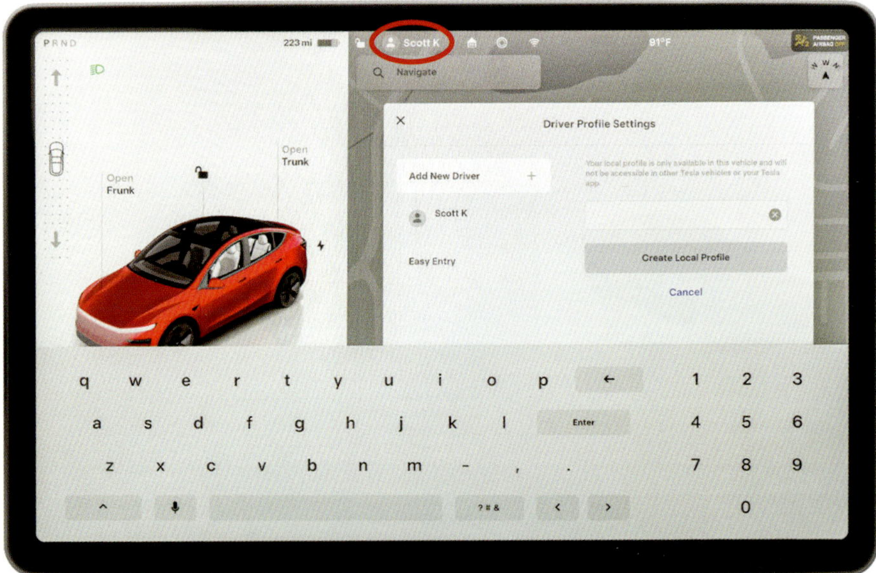

You save all these things to one place: your Driver Profile Settings. That way, not only are they saved, but they're tied to your phone key (and your key card), so when you're the one driving, it uses your settings. If your spouse or child is driving the car, you can set up additional profiles with their settings (more on this on the next page. BTW: When I say "child," you know I mean an older, licensed driver—not your 10-year-old). To set up your driver's profile, tap the Driver Profile icon (it looks like a person) in the Status Bar at the top of your center touchscreen to bring up the Driver Profile Settings window. Tap Add New Driver (near the top left), and then type in your name. Next, it's going to bring up a window where you can adjust your mirrors (see page 8) and seat (see page 6), as well as lots of other things you can customize for your most comfortable driving positions.

How Do I... Save Seat Position, Mirrors, etc., for a Second Driver?

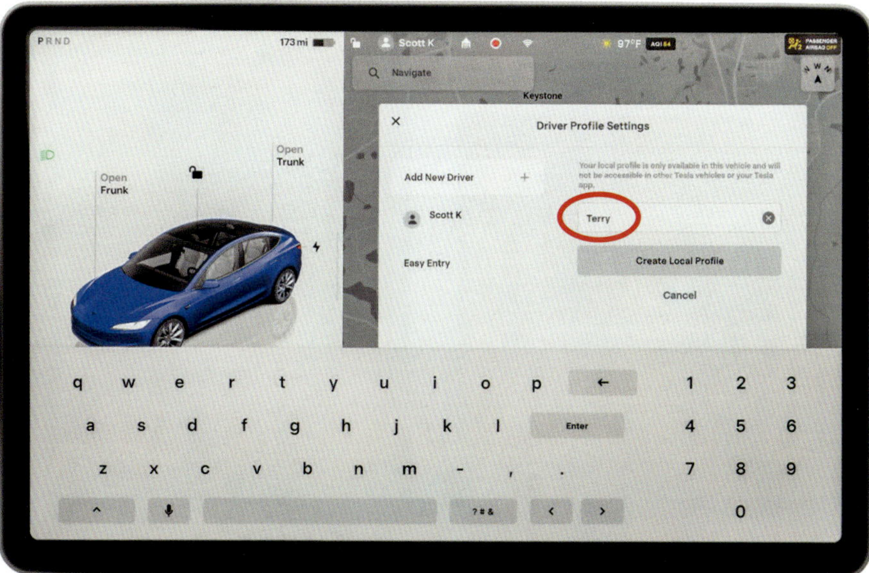

If you're going to have someone else drive your car on a regular basis (for example, I often wind up driving my wife's Model Y on road trips), you can set them up with a driver's profile with their own custom settings and options, including seat and mirror positions, so it's just one tap away. Here's how: Tap the Driver Profile icon (it looks like a person) in the Status Bar at the top of your center touchscreen to bring up the Driver Profile Settings window. Tap Add New Driver (near the top left), and then give this driver a name (I named mine "Terry"). It's then going to bring up a window where they can adjust the mirrors (see page 8) and seat (see page 6), as well as lots of other things they can customize for their most comfortable driving positions. For more on how to switch between driver's profiles, see page 45.

How Do I... Start My Tesla?

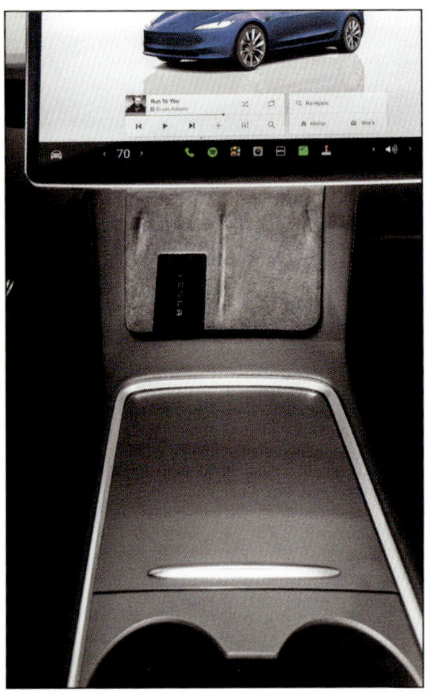

Once you've unlocked your car and you're sitting inside it, press the brake pedal, which will start everything that's not already running from you unlocking the car, and you're ready to put the car in drive and go. Now, what if you get in, put your foot on the brake pedal, and it's not ready to go? In that case, it will tell you on your center touchscreen to put your key card on either phone dock (beneath your touchscreen) or right behind the cup holders (in a Model 3 or older Model Y), and that will turn on all the driving stuff for you. It won't have you do this very often, but from time to time, for no particular rhyme or reason, it will (well, it does it to me anyway, and my wife, and my daughter, and another reason why, like I mentioned on page 3, I recommend keeping a key card in your wallet or purse). So, just know it's nothing you did—just place your key card there for a second or two and you'll be ready to go.

How Do I... Use a Regular Key Fob Instead?

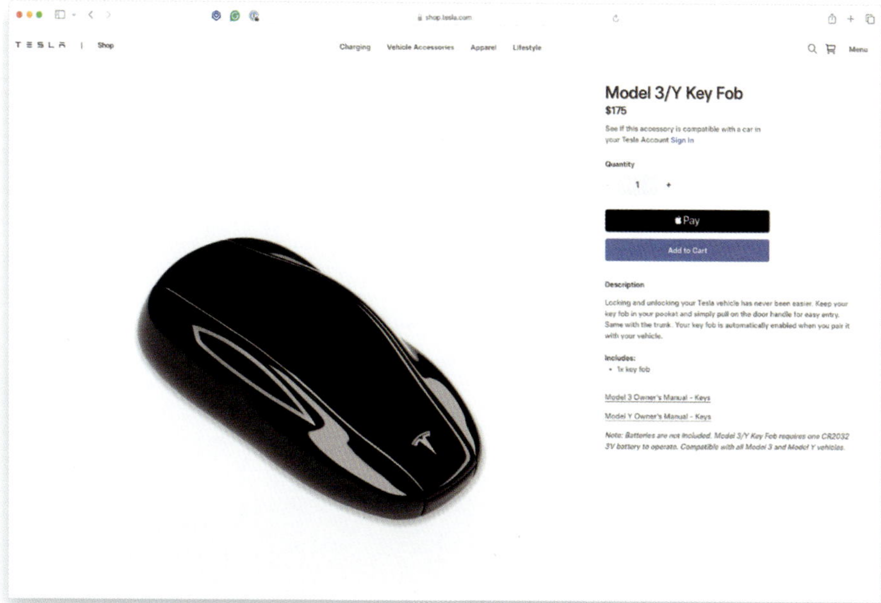

If instead of using your key card, you'd feel more comfortable using a traditional key fob (one you can put on a keychain) like our ancestors did (sorry, I couldn't resist), you can order one directly from Tesla.com. (*Note:* They are not cheap at $175. But, I have to admit—they're pretty cute.) It looks like a little version of your car, and it works pretty much like a phone key in that if the key fob is in your pocket or purse, it unlocks your door or trunk for you as soon as you touch your car's handle.

How Do I... Put My Tesla into Drive (or Reverse or Neutral)?

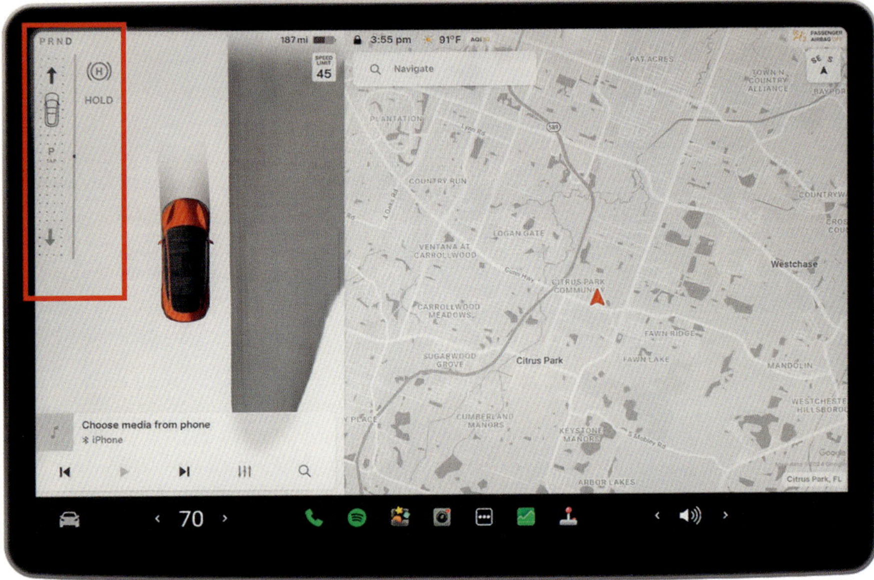

If you have a newer Tesla, there is no gear selector stalk or center shifter like in almost every other car in the world. Instead, you put your car into drive in one of two ways: (1) There is a Drive Mode Strip on your center touchscreen, and you can just swipe up on it with your finger to put your car into Drive (or down for Reverse). Or, (2) in the overhead console (in a Model 3 or newer Model Y. Not on the touchscreen—literally above your head, right above the rearview mirror) or in the center console (in a Model S or X), there is what Tesla calls a "Secondary Drive Mode Selector" where you can tap on the letter "D" to shift into Drive (or R for Reverse, N for Neutral, or P for Park). Both of these methods are weird. I know. The idea here is that your Tesla is supposed to know if you want to go into Drive or Reverse, and it will switch it for you automatically. For example, if you're in your garage and you need to back out, its cameras know you want to shift into Reverse, so it will do that for you once you step on the brake. That's the plan anyway (see the next page for more on this). If you have an older Model 3 or Model Y, then you'll use the gear selector stalk behind and to the right of your steering wheel, moving it up or down to switch gears. You can see which gear you're shifting into by looking at the top-left side of your touchscreen.

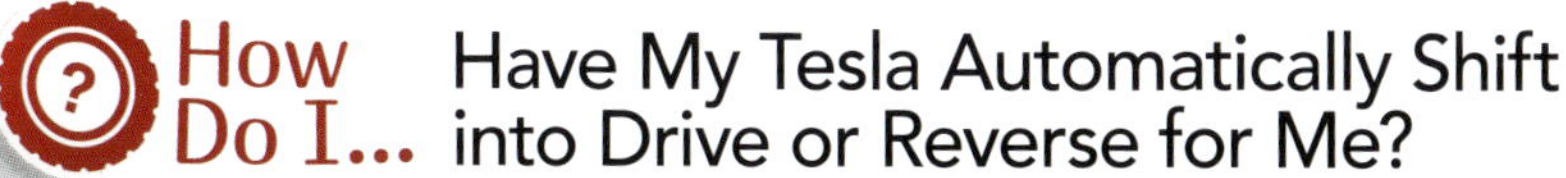

How Do I… Have My Tesla Automatically Shift into Drive or Reverse for Me?

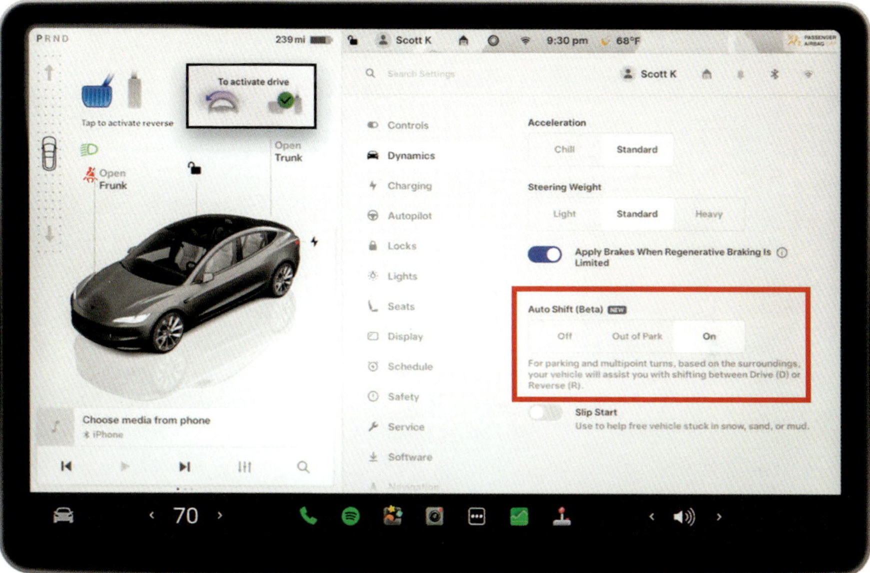

As I mentioned on the previous page, if you have a newer Tesla, there is no gear selector stalk. Instead, you use the controls on your center touchscreen (or in your overhead or center console, depending on your model) to shift into Drive, Reverse, etc. However, there's a cool feature called "Auto Shift (Beta)" where your Tesla automatically figures out which direction you want to go, and it shifts into Drive or Reverse for you (it's still considered a Beta feature at the time of this writing, meaning they are still testing it, but you can use it knowing it will get better over time). It works surprisingly well and it pretty much means you don't have to mess with shifting into Drive or Reverse at all—for the most part. The only time you have to touch your Drive Mode Selector is when you're done driving and want to shift into Park. You turn this feature on by tapping on the Controls icon (the car in the bottom left of your touchscreen), then tapping on Dynamics (or Pedals & Steering, depending on your model), and then tapping the Auto Shift (Beta) On button. Once it's on, this feature will only work when your seatbelt is fastened and your foot is off the brake. Here's an example of how it works: When I get into my car in my garage (seatbelt on, car started, and my foot off the brake), on my touchscreen it says, "Tap to activate reverse" below an image of a blue brake pedal. I tap the brake pedal, back out of my garage, and angle toward the street. When I stop, as long as I move the steering wheel a little, it automatically shifts into Drive for me, and away I go, never touching my touchscreen. Don't worry—it works better than it sounds, and you won't miss having to shift if it does it all for you.

How Do I... Put My Tesla in Park?

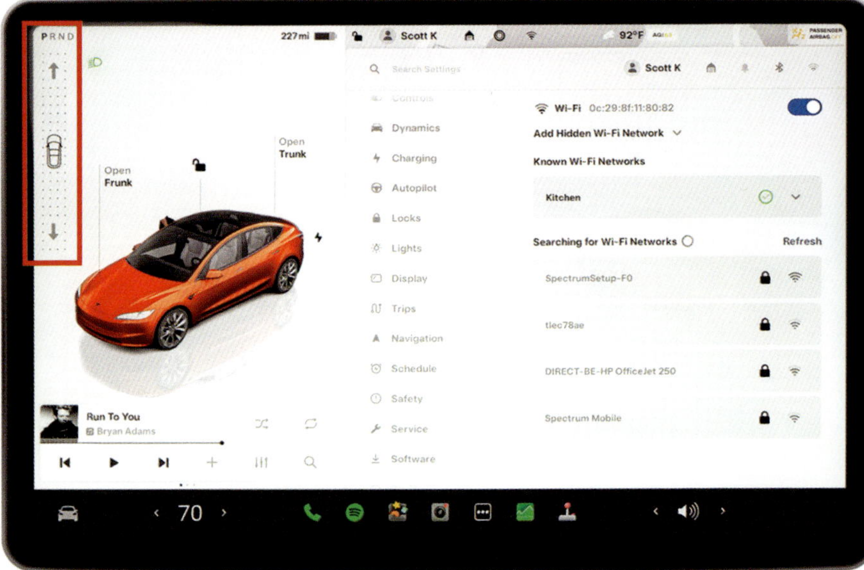

Newer Model 3 or Y (you have two choices here):

(1) Put your foot on the brake to bring up the onscreen shifter, and then tap the "P" in the top-left corner of your center touchscreen.

(2) In the overhead console (right above the rearview mirror), you'll see buttons for Park, Reverse, Neutral, and Drive—just press P for Park.

Model S or X (you have two choices here):

(1) Put your foot on the brake to bring up the onscreen shifter, and then tap the "P" in the top-left corner of your center touchscreen.

(2) Just under the wireless phone chargers near the top of the center console, you'll see a row of Shift controls with buttons for Park, Reverse, Neutral, and Drive—just press P for Park.

Older Model 3 or Y: There's a Park button at the end of the drive stalk behind the right side of the steering wheel. Press it once to put your Tesla in Park.

TIP: QUICK WAY TO SHIFT INTO PARK

When you unbuckle your seatbelt or open your door, your Tesla automatically shifts into Park.

How Do I... Use My Turn Signals?

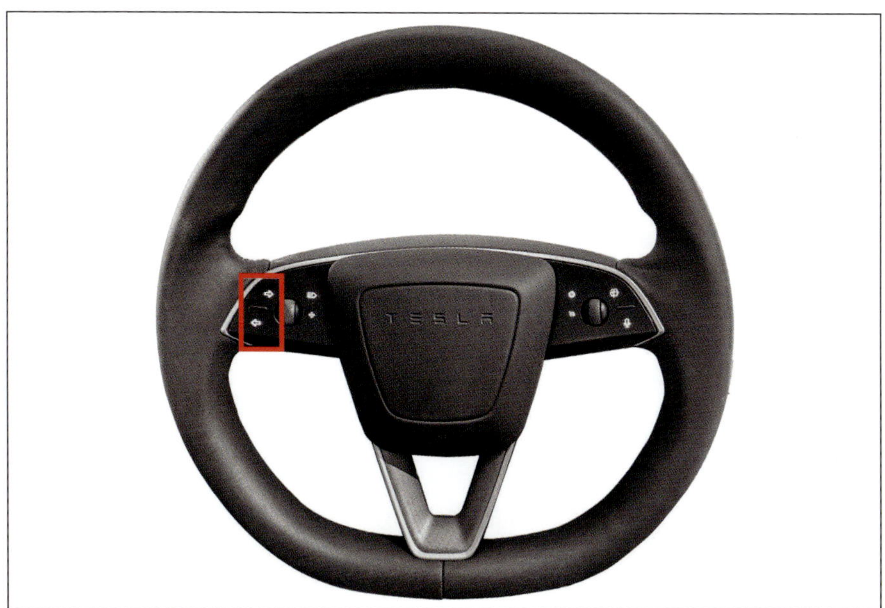

Model S, X, or newer Model 3 or Y: You'll use the left and right arrow buttons on the left side of the steering wheel itself (as seen above). If doing it this way feels weird at first, it's only because it is.

Older Model 3 or Y: It's the standard ol' turn signal stalk behind the left side of the steering wheel—pretty much the location for turn signals since the late 1800s.

Whichever way you engage it, you'll hear the familiar turn signal sound, and you'll see the turn direction arrow appear on your center touchscreen.

TIP: EASILY SEE YOUR BLIND SPOTS

When you turn on your blinker, you can have a live video appear on your center touchscreen, which shows you what the camera on that side of your car is seeing. This will help you see the lane you're moving/turning into, and if it detects a vehicle there, you'll see the side of the video turn red. To turn this on, tap on the Controls icon (the car in the bottom left of your touchscreen), then tap on Safety, and then turn on Automatic Blind Spot Camera. Once it's on, you can change the location of where this video feed appears by dragging-and-dropping it on your center touchscreen. (*Note:* See page 245 for more on this feature.)

How Do I... Have My Tesla Automatically Turn Off My Turn Signal When I Change Lanes?

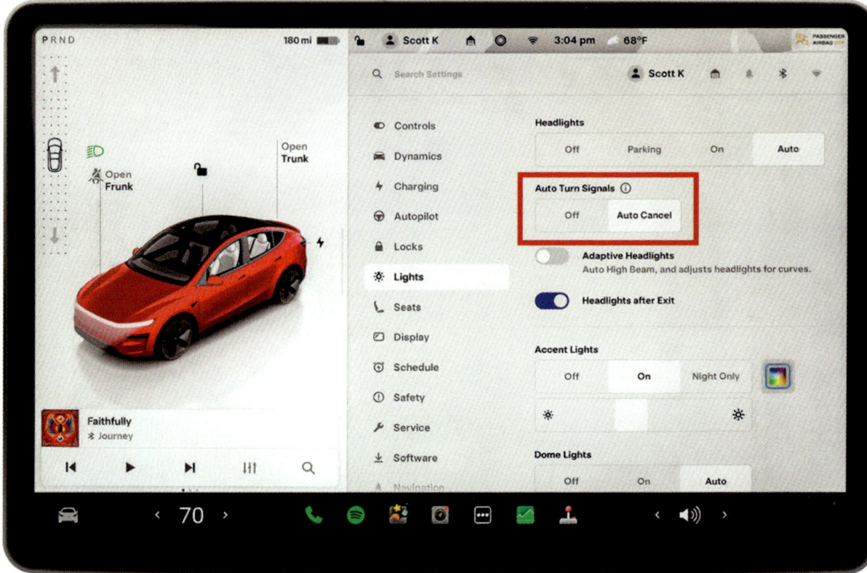

When you make a turn, once it's complete, your Tesla will, of course, turn off your turn signal for you (so you're not that goober driving down the road with your blinker on), but what if you're just changing lanes or merging in traffic? Well, there's a feature that uses your car's cameras to know when you've made the lane change or you've merged, and it will turn your turn signal off for you. To turn this feature on, tap on the Controls icon (the car in the bottom left of your center touchscreen), then tap on Lights, and then, under Auto Turn Signals, tap on the Auto Cancel button.

How Do I... Get an Alert When the Traffic Light Turns Green?

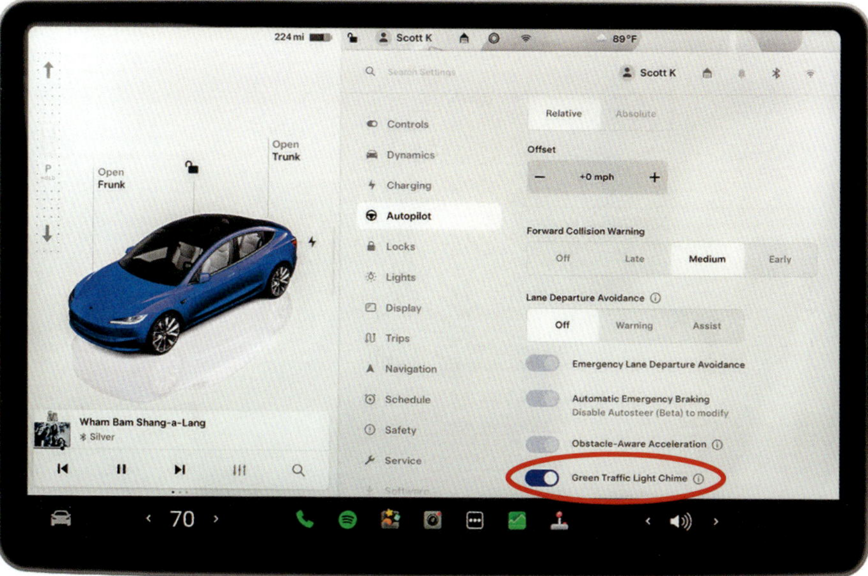

If you'd like a little chime to go off when you're stopped at a traffic light and the light turns green (yes, your Tesla recognizes traffic light colors), tap on the Controls icon (the car in the bottom left of your touchscreen), then tap on Autopilot (this feature is built in—you don't need Full Self-Driving to use this). Turn on the toggle switch for Green Traffic Light Chime, and son-of-a-gun, when that traffic light turns green, you can put your phone down and start driving again.

How Do I... Turn On My Headlights?

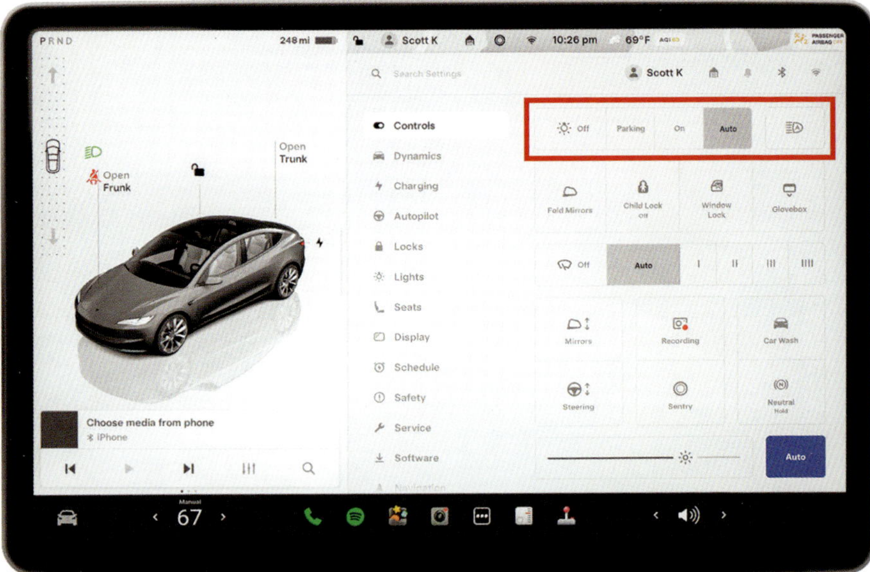

By default, they're set to automatically come on/turn off based on outside conditions, so in this Auto mode, you really don't have to worry about your headlights, parking lights, and stuff like that. But, of course, you can manually turn them on/off (or even disengage the whole Auto headlights feature) by tapping on the Controls icon (the car in the bottom left of your center touchscreen) and you'll see the options at the top of the screen. To turn your headlights off, just tap on Off (I didn't really have to write that one, did I?). If you tap on Parking, it turns your headlights off and just leaves on your parking lights, taillights, and other related lights (this is the mode I use if I'm acting as a getaway driver during a bank heist). Lastly, tapping On just turns your regular low beam headlights on. (*Note:* You'll also find these options by tapping on Lights on the Controls screen.)

TIP: QUICKLY ACCESS THE LIGHT SETTINGS MENU

If you want to use the standard "on, off, parking lights, etc.," type of controls, don't go digging through the Controls screen. Instead, just flash your high beams a couple of times (see the next page) and the Light Settings card pops up at the bottom of your touchscreen with controls for your lights.

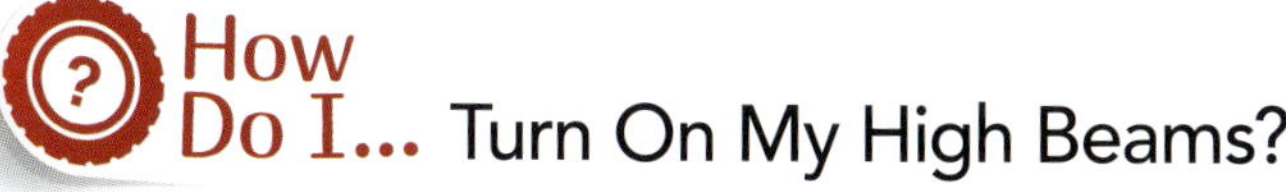

How Do I... Turn On My High Beams?

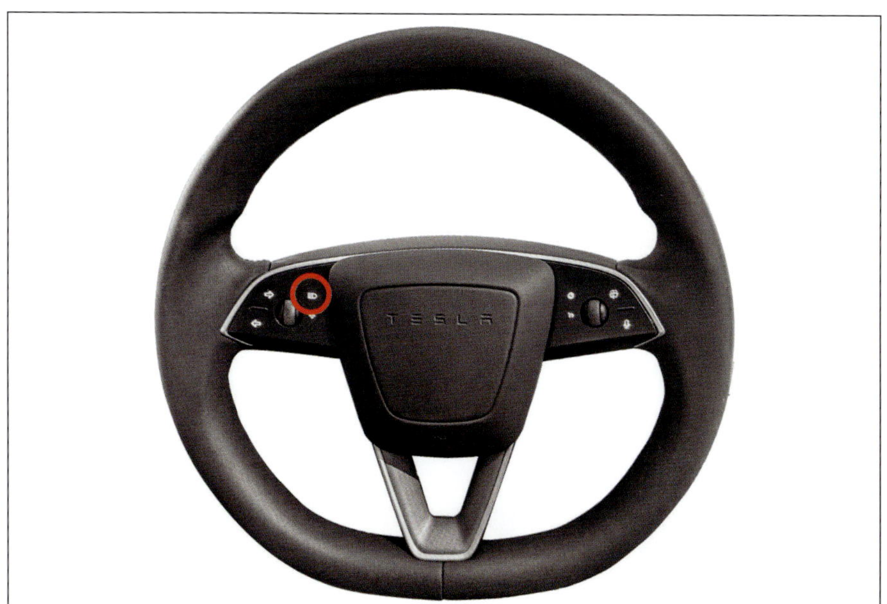

There's an automatic way and a manual way. If you turn on the automatic feature, your high beams will automatically turn on any time your Tesla thinks you need them. If it sees oncoming cars, it will automatically turn them off, returning to the regular low-beam headlights until those cars have passed. You turn this feature on/off by tapping the Controls icon (the car in the bottom left of your center touchscreen), then tapping Lights, and then tapping Adaptive Headlights.

The manual method depends on your Tesla model and when you bought it:

Model S, X, or newer Model 3 or Y: There's a High Beam Headlight button on the left side of the steering wheel—quickly press-and-release it to flash your high beams or press-and-hold it to keep them on. (*Note:* A little timer appears on your touchscreen [or instrument panel] to show how long you have to hold that button down to keep the high beams turned on. It's just a few seconds.)

Older Model 3 or Y: Pull the turn signal stalk behind the left side of the steering wheel toward you, and then release it to flash your high beams (like every other car you've ever owned). To leave them on, push the stalk forward. You'll see a blue high beams icon on your touchscreen letting you know they're on.

How Do I... Use My Windshield Wipers?

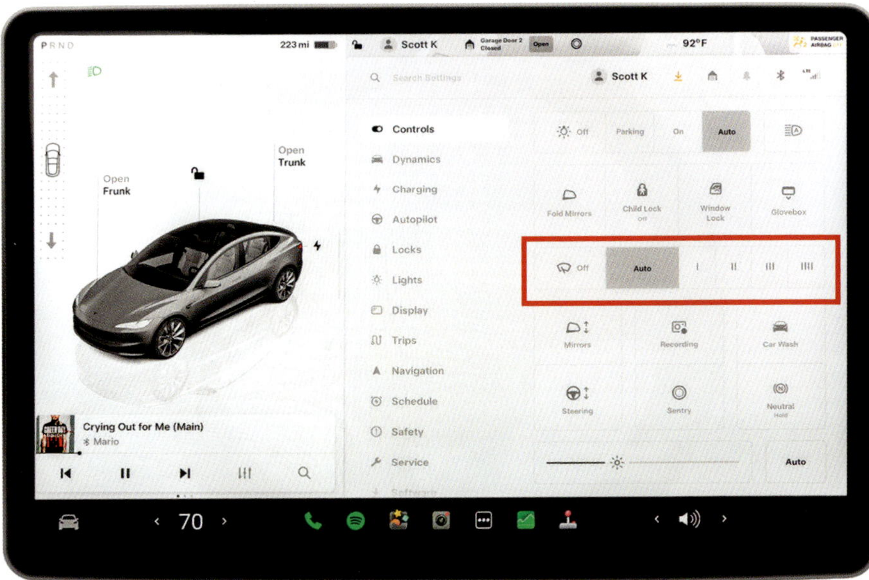

By default, they're set to Auto, so if your Tesla senses rain, it will automatically turn them on for you. If there's a little bit of rain, it'll do an intermittent style of wiping, and if it's raining hard, it will raise the speed of the wiping accordingly.

Model S, X, or newer Model 3 or Y: Press the Wiper button on the right side of the steering wheel and it does a quick wipe (in an older Model 3 or Y, press the button on the end of the turn signal stalk). If you want it to keep the wipers on, press the button again and a pop-up controller appears on your center touchscreen, but you can also just use the left scroll button on your steering wheel to change speeds or you can press the Wiper button multiple times to make them go faster. Press the button twice to go faster, three times for even faster, or four times for warp speed. If you press it a fifth time, it moves the speed back two to just "even faster," and if you press it again, it goes down to double-time, and so on.

TIP: TELL YOUR TESLA TO TURN ON/OFF YOUR WIPERS

You can also use a voice command anytime to control the wipers by pressing the Microphone button (or the right scroll button in older models) on the right side of your steering wheel, then releasing it and just saying what you want to do. For example, "Wipers off" or "Wipers on."

How Do I... Wash My Windshield?

Model S, X, or newer Model 3 or Y: Press-and-hold the Wiper button on the right side of the steering wheel and it sprays wiper fluid onto your windshield. When you let go of the button, it does a few wipes to remove the fluid. It will spray wiper fluid for as long as you hold that button.

Older Model 3 or Y: Press-and-hold the button on the end of the turn signal stalk behind the left side of the steering wheel.

How Do I... Use My Brake Way Less?

Pre-2024 Teslas had three different types of braking, called "Stopping Modes," which were Hold, Creep, and Roll. Now, the Hold stopping mode is baked in for you, so there's nothing to choose or turn on—it's the default mode. I think Hold is far and away the best anyway, as it uses Tesla's regenerative braking—when you take your foot off the gas, instead of coasting (like a gas-powered car), your car starts slowing down like you're smoothly braking. A lot of folks fell in love with this feature (me being one of them) and I bet you'll probably find you use the actual brake pedal just a fraction of what you would in a gas-powered car. That's probably why Tesla made it the default. The other cool thing about Hold stopping mode is while your foot is off the gas, and the car is gently braking, it feeds energy back to the main battery, so it helps extend your range. Also, when you're at a stop light and take your foot off the brake, your car won't creep ahead like gas-powered automatic transmission cars do—it will just hold there for you.

How Do I... Turn on My Parking Brake?

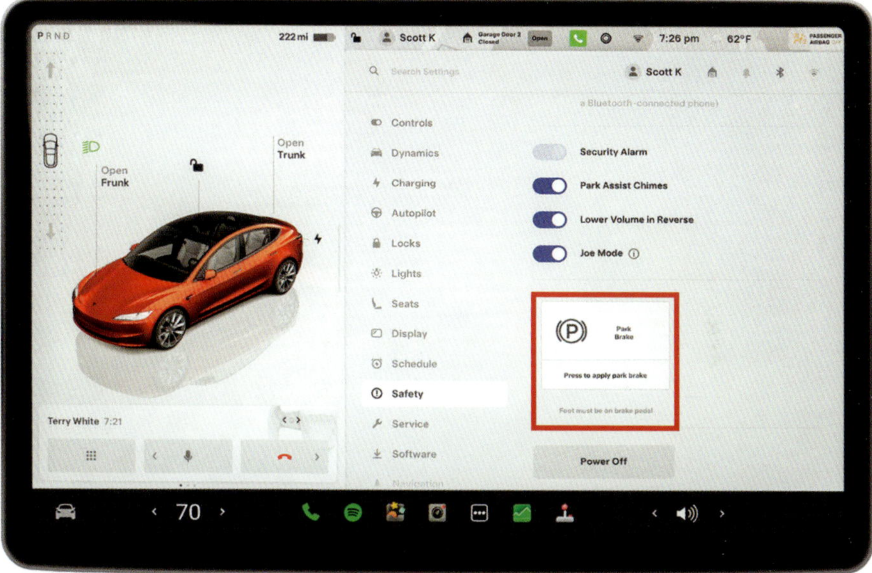

Model S, X, or newer Model 3 or Y (you have two choices):

(1) You can go to your center touchscreen, tap on Controls (the car icon in the bottom left), then on Safety, and while putting your foot on the brake (the button won't appear without doing this), tap on the Park Brake button.

(2) But, it's faster to just tap-and-and hold the "P" on the top-left corner of your center touchscreen. To disengage it, just shift to Drive or Reverse (or even Neutral).

Older Model 3 or Y (you have two choices):

(1) You can go to your center touchscreen, tap on Controls (the car icon in the bottom left), then on Safety, and while putting your foot on the brake (the button won't appear without doing this), tap on the Park Brake button.

(2) But, it's faster to just press-and-hold the Park button at the end of the drive stalk behind the right side of the steering wheel for a few seconds. You'll see a symbol with the word "PARK" in red letters appear on your touch-screen, letting you know the parking brake is engaged. To disengage it, just shift to Drive or Reverse (or even Neutral).

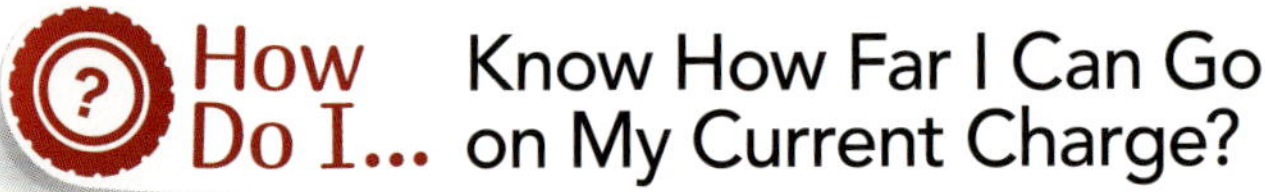

How Do I... Know How Far I Can Go on My Current Charge?

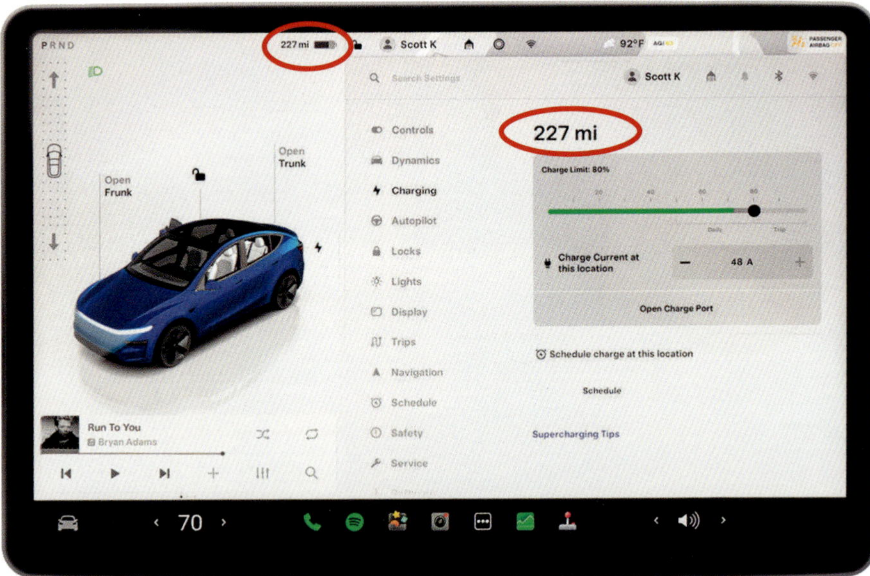

There are a couple of different ways to find out. If you're not in your car, you can launch the Tesla app on your phone and right at the top of the screen, in big numbers, it will display how many miles you can go. It's fairly accurate as long as you don't go zooming all over the place at high speeds. It's based on you driving 55 mph (like standard EPA estimates) or slower, so if you go zooming around at 80 mph, it uses up more juice (just like a gas car would use up more gas if you drove it at 80 mph). But still, it's reasonably accurate. Once you're in your car, you can glance at your center touchscreen and in the top-left third of it, to the right of your current gear (like D or R), it will show the estimated range based on your current charge (it's the same number you'd see in the app). If you want to "nerd out" on your charge stats, tap the battery icon to the right of current charge and it'll give you more stats than you probably want. You can also nerd out a bit in the app on your phone by tapping on Charge.

How Do I... Know If I'm Running Out of Battery Power?

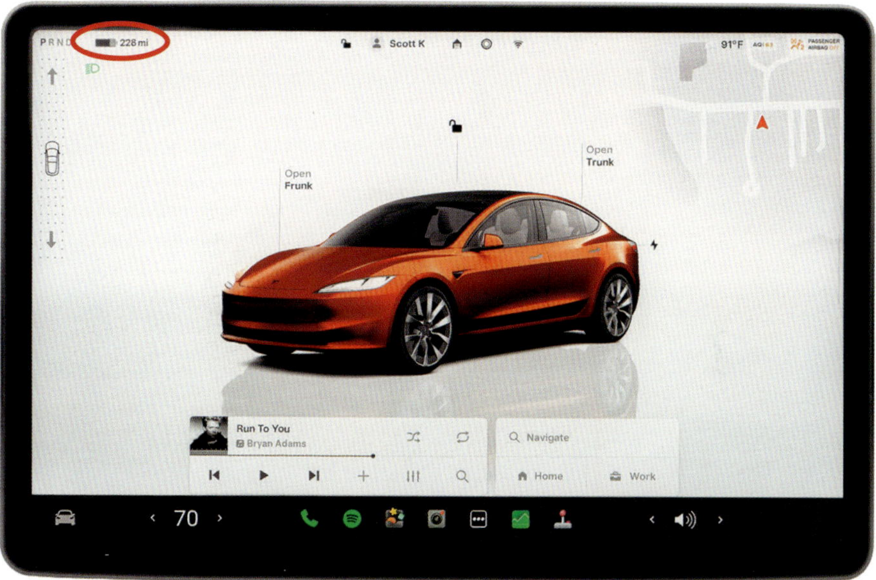

Your car will tell you. It will start to bring up warnings when you get a bit too low, and the warnings get more serious the lower you go. Your Tesla displays your battery level right onscreen, and the battery icon at the top of your center touchscreen should stay gray or green, but if you fall below a 20% charge, you'll see it turn yellow, warning you it's probably time to charge up. If you get much lower, you'll get an onscreen warning, especially if you're far away from a known charger (like your house or a Supercharger). It will tell you "Hey, you're almost out of range of a known charger," so you can navigate to a nearby one. One big way it protects you from running out of battery power is if you enter an address or destination in the navigation system. If that destination is farther than your battery charge will take you, it will automatically insert Supercharger stops along your route, and it even lets you know how long you'll need to charge there to make it to your destination with some battery power to spare. To help you find the nearest charging station anytime, on the map, tap the lightning bolt icon and it will highlight Superchargers and other public charging stations, like ChargePoint. If your battery power gets really low, you'll see a small orange Turtle icon appear and your Tesla will start turning off features that eat up a lot of battery power to help you get to a charging station (or back home to your charger), and it will limit your speed to save your battery power as well.

How Do I... See How Long My Current Trip Has Taken?

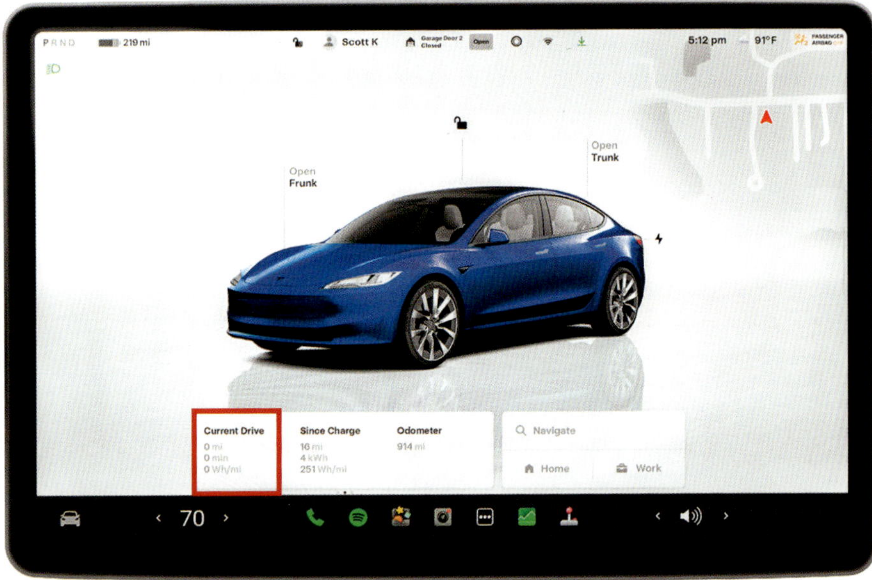

Swipe to the right just below the illustration of your car on the left side of your center touchscreen and the shortcut "cards" (mini-panels) appear with your current trip info, including how long it has been since you started your car, how many miles you've gone, and how long it has been since the last time you charged. If you want more detailed information (including setting up your own custom trip timers), tap on the Controls icon (the car icon in the bottom left of your touchscreen), then tap Trips, and you'll find more info on Energy and the ability to create your own custom trip counters.

TIP: SEE HOW MUCH ENERGY YOUR TESLA IS USING

Electric cars are a lot like gas cars in that they use up fuel (or electricity) in very similar ways. For example, in a gas car, if you drive fast, it uses up more fuel, and if you drive fast in a Tesla, it eats up more energy. The air-conditioning uses more fuel in a gas car; same thing in an electric car. Well, if you find this type of stuff fascinating, have I got an app for you: tap on the App Launcher icon (the three dots) at the bottom of your center touchscreen, and from the app tray that pops up, tap on Energy. You're now in nerd heaven.

How Do I... Plug In the Charging Cable?

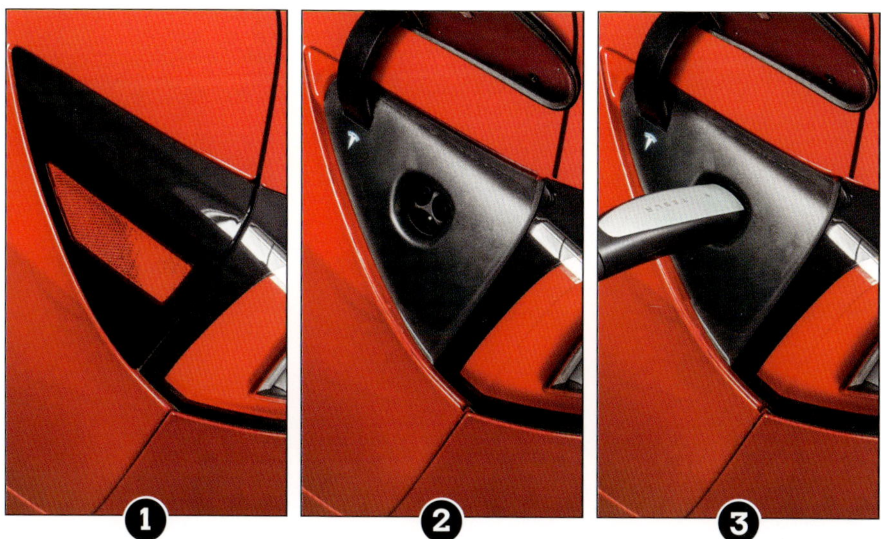

Walk to the back driver's side of your car and you'll see some side brake lights and reflectors, but the large one (#1 above) cleverly hides your Charge Port. Once you find the port and unlock it, the rest is pretty much like charging your phone—plug it in and it charges. It's that easy. To unlock the Charge Port, press the round button on the top of the charging cable and the door pops open revealing the Charge Port (#2). Now, just plug in the charger (#3). If it's securely in, you'll see the small Tesla logo to the left of the Charge Port start to flash blue, then green, and all is good. If for any reason it doesn't flash either of those (hey, it happens), press the button on top of the charger again to release it, pull it out, and then try plugging it in again, making sure it's firmly in place. Of course, this is a Tesla, so there are more ways to open the Charge Port: On the left side of your center touchscreen, you'll see a 3D illustration of your car. Tap on the little lightning bolt icon near the rear driver's side and that pops the Charge Port open. You can also use a voice command (see page 36) by simply saying, "Open Charge Port," and that will do it for you, too. Lastly, you can go to the Tesla app on your phone, tap on the Charging icon (it looks like a little lightning bolt), and then in the Charging menu that appears below it, tap on Unlock Charge Port, and that will open it, as well. When you unplug the charger and walk away from your car, the Charge Port will close automatically. You can also tap on the little lightning bolt icon near the driver's side of the 3D illustration of your car on your touchscreen and that'll also close it for you.

How Do I... Open My Garage Door from My Tesla? (Option #1)

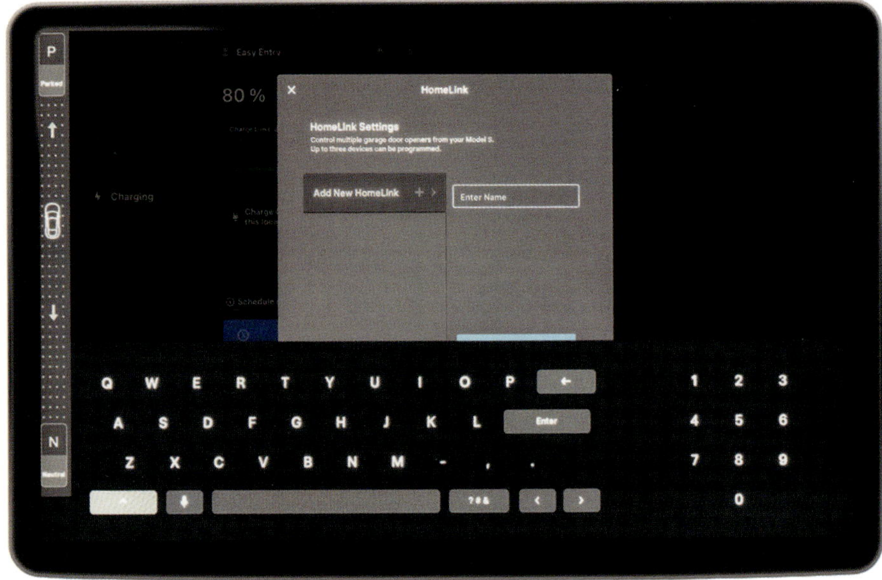

If you have a newer Tesla, it doesn't come with the built-in HomeLink feature for opening your garage door or (if you have one) the gate to your neighborhood (it used to come standard, but not anymore—don't get me started). So, you'll have to (a) order the HomeLink feature separately ($350, last time I checked), but this is not a do-it-yourself install. You'll have to (b) schedule Tesla to come out and install it for you (which somehow, they don't charge you for), or if they can't come out, you'll have to schedule a time to take your car into the service center for the install (don't forget to take the unit you ordered with you). Once it's installed, you'll have to pair your garage door with your car, which is both easy and a bit of a pain because chances are you'll have to pull your car out of the garage, take out a ladder, open the rear cover of your garage door opener, and hit the "learn" button, so it emits a signal your Tesla can read. After four or five tries (following the step-by-step instructions in your Tesla, under the HomeLink section on the center touchscreen), it will finally work, and then when you come home (or you're ready to leave), you can tap the HomeLink button at the top of your center touchscreen, tap the Activate button, and up/down goes the garage door like magic. Just know, this setup process always seems to take multiple tries. It's entirely possible you won't need to go through all that, but I doubt it. Let's hope for the best-case scenario, which is the whole thing works the first time without you having to pull out a ladder or remove any covers. Now, have I ever heard of that scenario playing out? No. Not once. Could it happen? Well, anything can happen. Is it likely? Ummm...no. The good news is, once it's done, it's done, and you won't have to do it again.

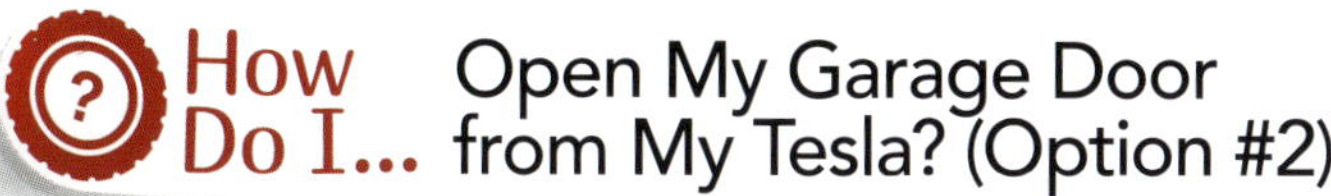

How Do I... Open My Garage Door from My Tesla? (Option #2)

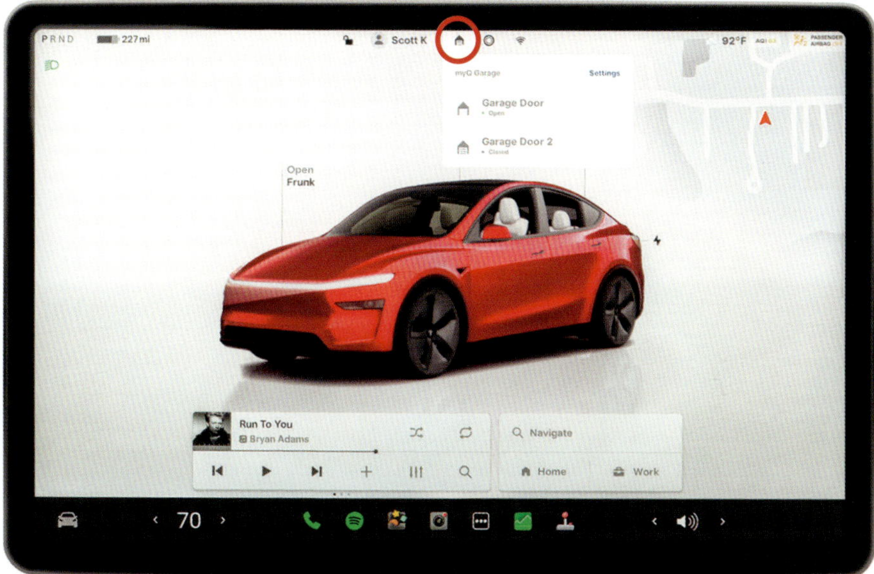

The other option for opening your garage door from within your Tesla is to have a garage door opener that has a built-in app that can integrate with your Tesla (or, buy an adapter that integrates with your current garage door opener—more on that in a minute). For example, my house came with a LiftMaster brand garage door opener and it uses LiftMaster's myQ app, which integrates with a Tesla, so I can open/close my garage door from my center touchscreen controls without actually having HomeLink installed. For this to work, you have to pay a monthly subscription fee to LiftMaster, and the last time I checked it was $99 per year (so, around $8 a month), so opening your garage door from your car is going to cost you one way or another, but at least it's an option, right? Now, if you don't have a LiftMaster garage door opener, you can buy a separate myQ adapter to work with many existing garage door openers, so it's worth checking out if that's the case. But, it can be a bit of a pain in the @#$ to install it, unless you're pretty handy with installing, and drilling, and stuff (I'm not).

How Do I... Make Getting In/Out of My Tesla Easier?

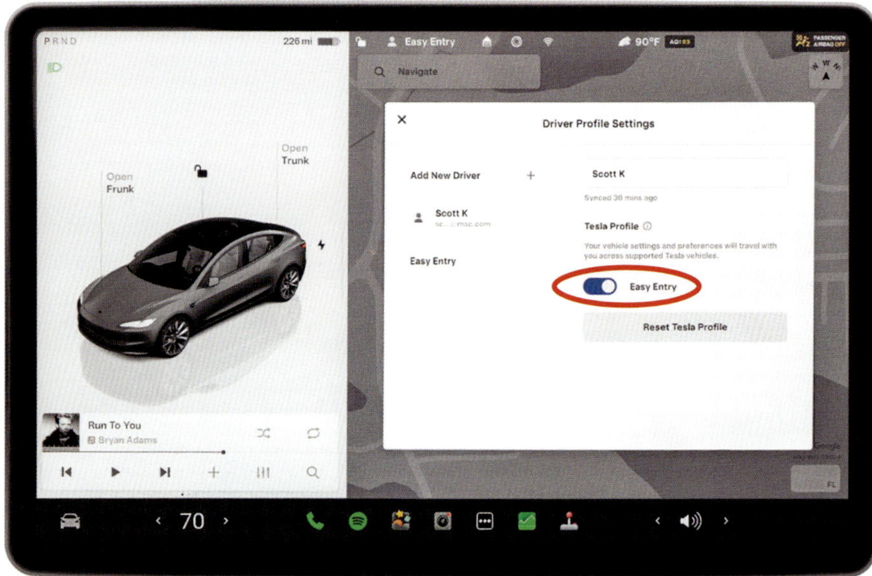

The Easy Entry feature automatically moves the driver's seat backward (and up higher if you want) and pushes the steering wheel back out of the way to make getting in and out of your car easier. If you turn this feature on, it kicks in once you're in Park and unbuckle your seatbelt. To turn on Easy Entry, tap on the Driver Profile (the person) icon in the Status Bar at the top of your center touchscreen to bring up the Driver Profile Settings window. Tap on your current driver's profile (if you don't have one yet, jump over to page 10), and then tap on Easy Entry. This brings up a window instructing you to move your steering wheel to the position where you want it to retract to (using the left scroll button on your steering wheel), and then use the controls on the left side of your seat to move your seat all the way backward (or however far back you want it). Lastly, tap the Save button to save Easy Entry to your driver's profile. Now (and this is totally optional), you could also create a separate driver's profile and add Easy Entry to just that profile, so you could have one for when you want Easy Entry and one for when you don't. Just add the word "Easy" to the end of the profile name, so you'll know which one is which. Of course, you could just add an "E" or "EE" for Easy Entry—I'll leave the naming part up to you. But, again, the whole extra driver's profile thing is totally optional.

How Do I... Lock the Doors When I Get Out?

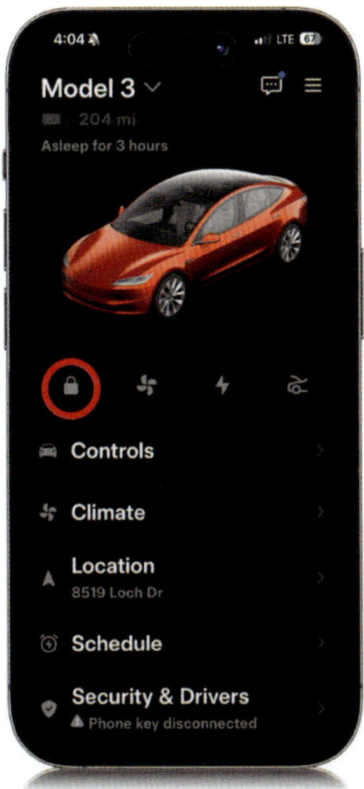

All you have to do is walk away. That's it. Just walk away and your Tesla will automatically shut down, fold in the mirrors, and lock all the doors and trunks for you. This takes some getting used to at first, and you'll keep looking back at your car to see if it really buttons up and locks, but you can relax—it works based on the location of your phone or key card and I've never had it not work and leave my car unlocked. Not once. (*Note:* If you'd feel better hearing an audible sound that lets you know that "Yes, your car is locked," see page 59.) If you want to do it manually, you can go to the Tesla app on your phone and tap the Lock icon that appears in the row of buttons right under the image of your car (its icon looks like a lock), but that's the hard way.

How Do I... Connect to My Home WiFi?

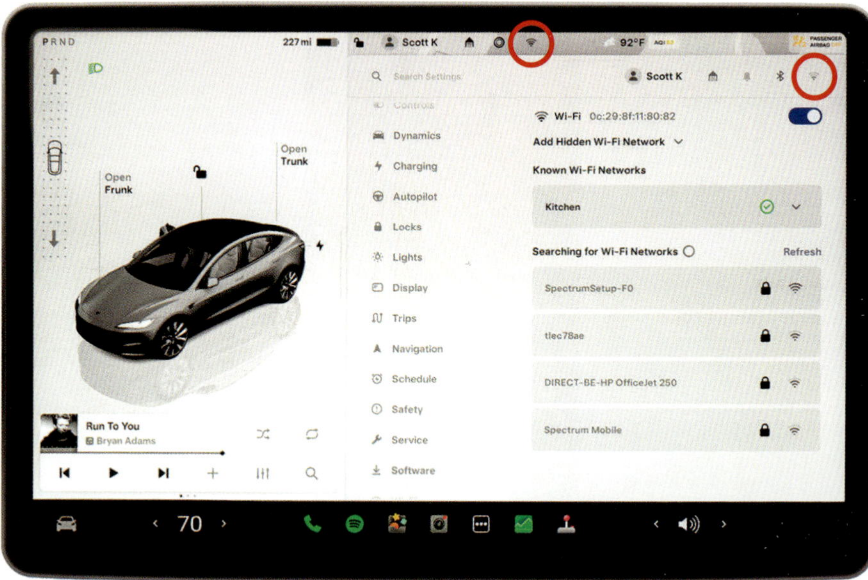

You probably already know by now that your Tesla gets new features, and bug fixes, and fun toys added to it via wireless software updates using your car's built-in cellular connection. However, if you want faster downloads for these updates, you can connect your Tesla to your home's WiFi network, and then everything moves faster than a greased pig (a phrase you don't get to hear every day). As long as your WiFi signal reaches your garage (or wherever you keep your Tesla), then all you have to do to connect it is go to your center touchscreen, tap on the Controls icon (the car in the bottom left), and up in the right corner of the Controls screen is a little LTE icon that shows your car's built-in cellular WiFi. Tap on that and your car will start searching for nearby WiFi signals. When you see yours, tap on it, and it will prompt you to enter your password. Enter it, then tap the Confirm button, and in a few seconds, you're connected to your home WiFi. Yes, it's as easy as that.

How Do I... Connect My Phone to My Tesla?

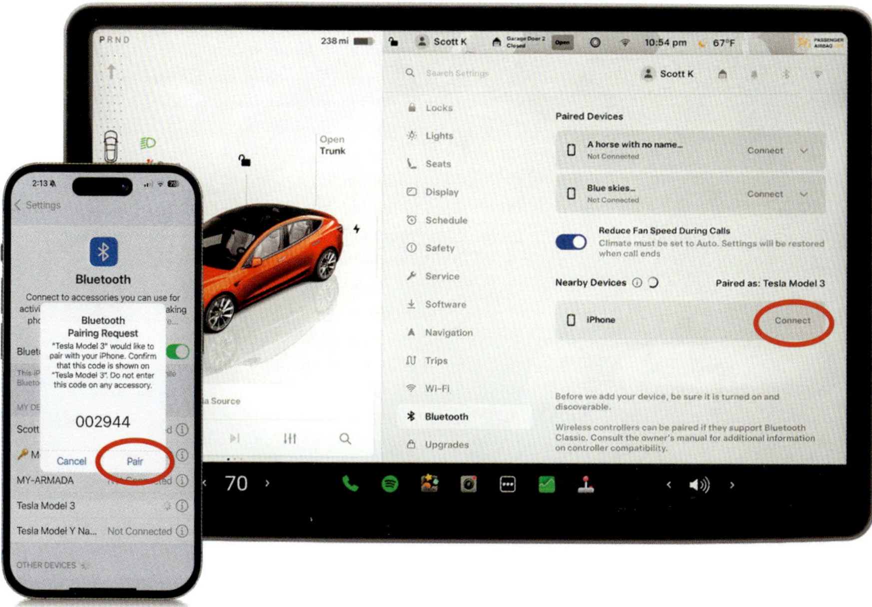

The first thing you need to do is go to your phone's Bluetooth controls and make sure it's discoverable (for example, on an iPhone, all you have to do is go to your Settings, tap on Bluetooth, and with it turned on, it puts your phone in discoverable mode). Next, in your Tesla, tap on the Controls icon (the car in the bottom left of your center touchscreen), then tap on Bluetooth to bring up its screen. At the top, you'll see a list of any paired Bluetooth devices that are connected to your Tesla, but since this is your first time connecting your phone (or anything else), you won't see a list at all. So, tap on Start Search in the bottom right. Since your phone is in discoverable mode, after a few seconds, you should see its name appear under Paired Devices (if you don't see it, double-check that it's in discoverable mode, or try restarting it, then go back to its Bluetooth screen to make it discoverable, and then try your Tesla's search again). Once you see your phone, tap Connect to the right of its name, and it will bring up a code onscreen and your phone should bring up that same code. If the two codes match, on your phone, tap on Pair, and your phone and Tesla are now connected. Now, on your phone, a pop-up will ask if you want to sync your phone's contacts, calendar, and media. Just click Allow (or Don't Allow—dealer's choice), but you can also do the same thing (syncing your contacts) right on your Tesla's Bluetooth screen, under the settings for your connected phone (it's a toggle switch). By the way, you'd do this same process to add your spouse's or kids' phones, or any other Bluetooth devices.

How Do I... Use Voice Commands to Do Stuff?

It's amazing how many features in your Tesla are available by simply verbally asking. To use a voice command, just press the Microphone button on the right side of the steering wheel once. You'll hear a chime sound and see a green circle with a microphone on it appear at the bottom left of your center touchscreen, letting you know your car is listening. Just tell it, in plain English, what you want it to do. "Navigate to Steak 'n' Shake," or "Turn up the fan," or "Open the glove box," or "Call Grandma," or "Play the song 'Goodbye Yellow Brick Road'," and so on. It actually does a really good job of understanding (it's not 100%, but certainly in the high 90s) and carrying out your commands, and I use this all day every day. If you have a Tesla made before 2024, then there is no Microphone button. Instead, you press the scroll button on the right side of the steering wheel to turn on voice command. Just press it once, then release and start commanding!

How Do I... Customize Which Apps Appear at the Bottom of My Touchscreen?

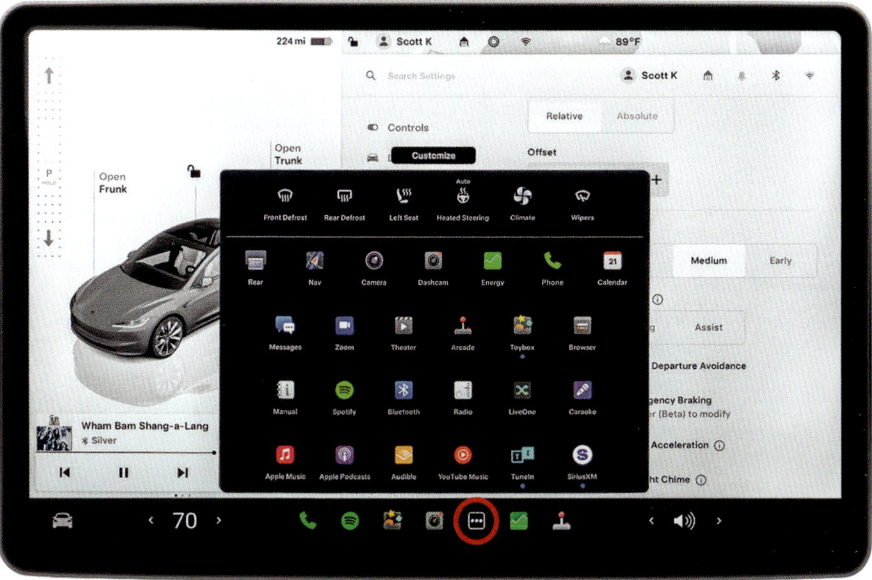

Tap on the App Launcher (the three dots) icon at the bottom of your center touchscreen, and from the app tray that appears, tap-and-hold on the app you'd like to appear in the touchscreen's bottom bar and simply drag-and-drop it right where you want it. You can have up to five of your favorite apps in the My Apps area of the bottom bar, to the left of the App Launcher. If you want to delete an app (so you can replace it with another), tap-and-hold on it, then tap on its little "X." Also, you'll see two more apps to the right of the App Launcher icon—those are your two most recently used apps.

How Do I… Change My Touchscreen from Light Mode to Dark Mode?

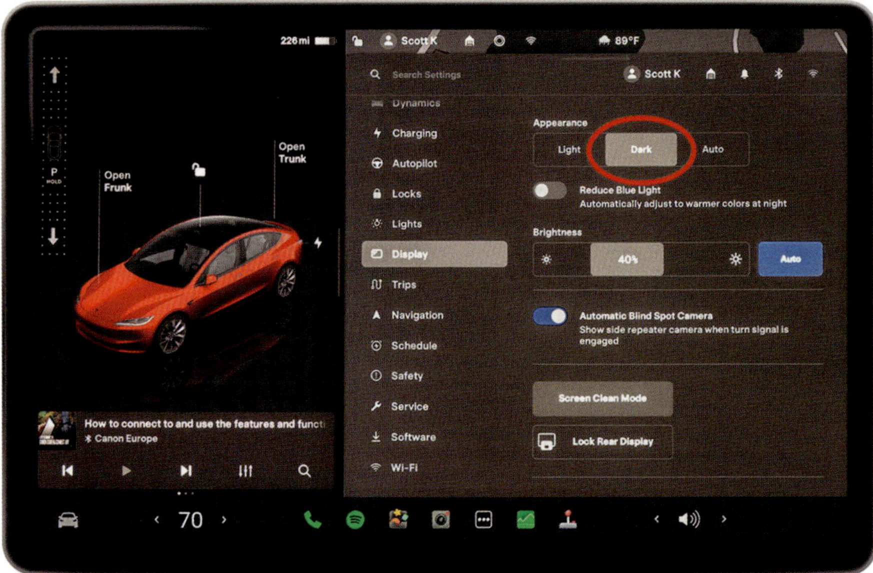

Tap on the Controls icon (the car in the bottom left of your center touchscreen), then tap on Display, and beneath Appearance, tap on Dark and your touchscreen background will turn black with your text reversed in white. If you choose Auto, it will automatically switch to Light (bright white) mode during the day, and when it turns to dusk, your touchscreen will automatically switch to Dark mode.

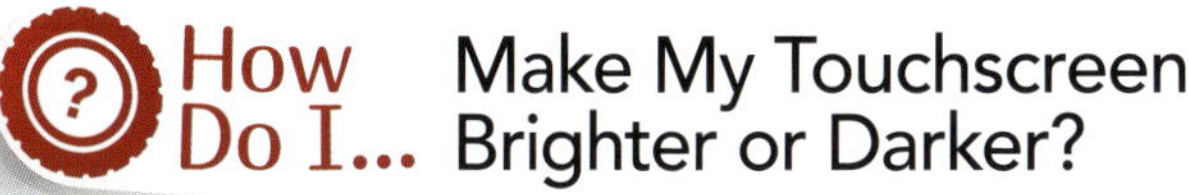

How Do I... Make My Touchscreen Brighter or Darker?

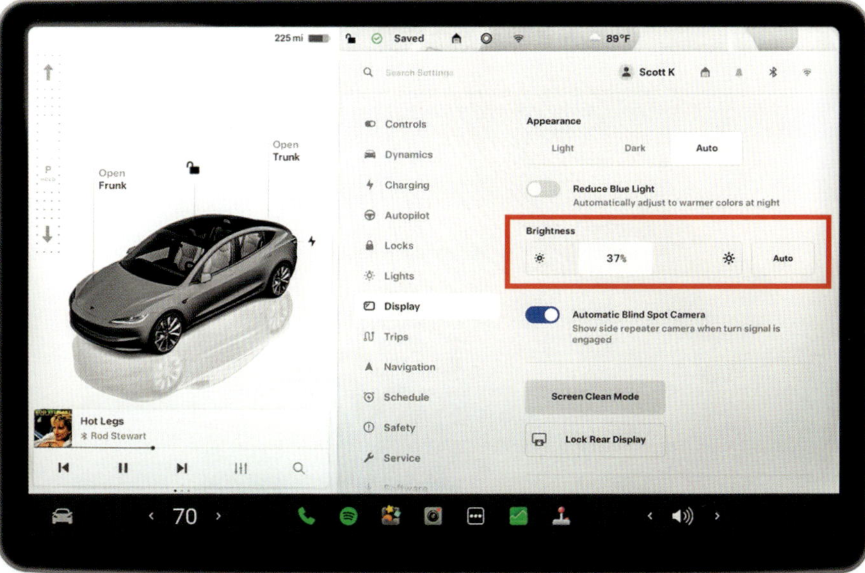

Tap on the Controls icon (the car in the bottom left of your center touchscreen), then tap on Display, and beneath Brightness, you'll see a slider. Just tap-and-drag that slider left or right to make your touchscreen darker or brighter. If you tap on the Auto button, your touchscreen will automatically adjust its brightness based on your current surroundings (so, it will automatically darken at night or on a bright sunny day, or brighten a bit on a very cloudy or rainy day, and so on).

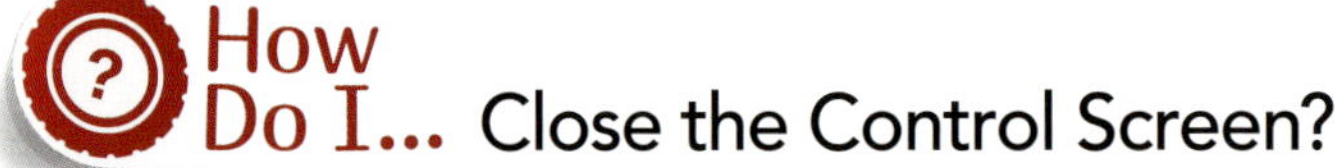

How Do I... Close the Control Screen?

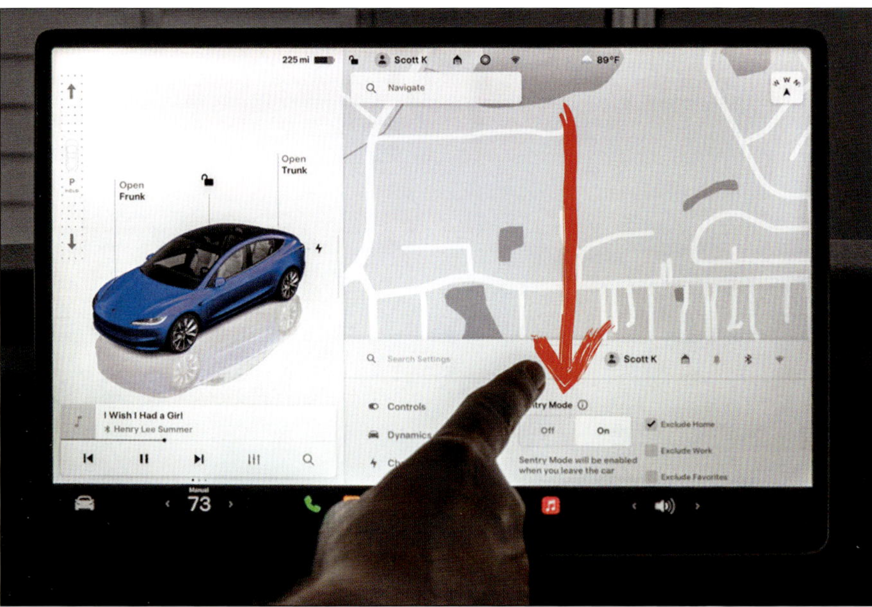

If you've opened an app or a window or the Media Player or pretty much anything other than the map, you can close that and return to the main screen (which is the map, by the way) by either tapping once on the Controls icon (the car in the bottom left of your center touchscreen) or simply swiping down on the currently open app or window to close it. I usually go with the Controls icon because (a) swiping doesn't work 100% the first time, and (b) sometimes I accidentally tap something else and launch it while swiping, so...yeah, there's that.

TIP: DISMISSING POP-UP ALERTS

Your Tesla communicates with you in real time with little pop-up alerts that appear in the bottom-left corner of your touchscreen (you'll see things like "Fasten your seat belts" or "Passenger door open"). If you want to get rid of any one of those popups when you see them onscreen, just swipe down right on the alert.

How Do I... Get a Better View to Clean My Touchscreen?

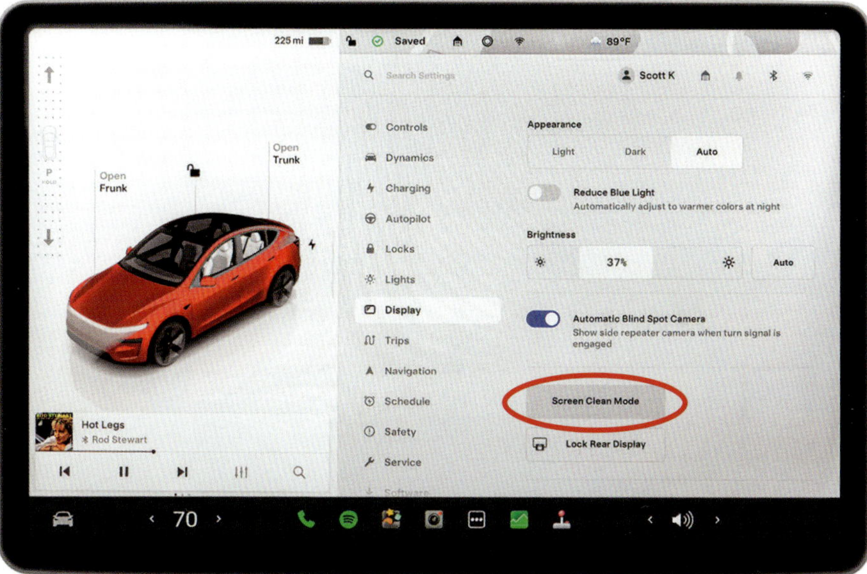

There is a special mode for cleaning your center touchscreen, and it's probably the easiest thing Tesla has ever done because all this mode does is turns off the touchscreen so it's just a black, empty screen where you can clearly see the finger prints and junk. That's it. This is the only thing on a Tesla I probably could have written the code for myself. Well, actually, probably not even that. I'm not a coder. Anyway, here's where you enter Screen Clean Mode: tap on the Controls icon (the car in the bottom left of your touchscreen), then tap on Display, and then tap on the Screen Clean Mode button. Now, the real trick is how do you get out of Screen Clean Mode since the entire screen is black. Well, it does leave one thing onscreen: a paragraph of text and a button that says "Hold to Exit." You hold that. To exit. I know. Duh. By the way, once you're in Screen Clean Mode, to clean the surface of your touchscreen, use a fresh, lint-free microfiber cleaning cloth (you can get them on Amazon), but don't use Windex or any screen cleaning solution. Tesla says don't even use a wet wipe. Just take the microfiber cloth—that's all it needs—and wipe off the touchscreen. When you're done, again, just use that Hold to Exit button.

Chapter 2

How to Do Everyday Driving Stuff

Let's Take Her for a Spin

Do you have any idea how many options and features you have access to in your Tesla's center touchscreen (remember, there aren't any buttons on the dash—just the handful of buttons on the steering wheel—so every other thing you might need to do in your Tesla is all in that one touchscreen). Do you give up? Great. I don't really know either (who has the time to count up stuff like that), but let me tell you this: it's a whole bunch. Now, since you're already accustomed to taking quizzes (thanks to the first chapter's intro), why don't we do another? This one is harder because I'm going to list a bunch of names of Tesla features and you have to pick which one isn't an actual feature. It sounds like this would be really easy, and it would be if you didn't realize that Tesla has some really "out there" kind of features. Grab your #2 pencil and let's begin. Q. Which of these is *not* an actual Tesla feature: (A) Boombox, (B) Mars, (C) Romance, (D) Oil Slick, (E) Toybox, (F) Emissions. Not sure, right? I know, it's tricky. I'll give you a few more seconds. Okay, pencils down. The correct answer is (D) Oil Slick (that feature is only found in James Bond's Aston Martin, which if you ask me is limiting its potential worldwide market). The rest are all names of real Tesla features (as you'll learn later in the book), but since you got that one right, you get a chance to play in our bonus round. Q. Which of these *are* names of actual Tesla options or features: (A) Rainbow Road, (B) Rebalance, (C) Rotate Tires, (D) Biofreeze, (E) Change Oil, (F) PlayJams. Okay, pencils down. The correct answer is (A) Rainbow Road (see page 126). I'll bet you chose either (E) Change Oil or (F) PlayJams, and that's what makes a question like this so challenging. To be honest, the deck was stacked against you from the start, as I used a well-known technique employed by diagnostic researchers in college-level academic testing, which is to use a word, like "Oil," which you are familiar with—from changing it in gas-powered cars—and, therefore, you think that's the answer. You may begin your lunch period. Class dismissed.

How Do I... Add an Additional Driver with Only Phone Access?

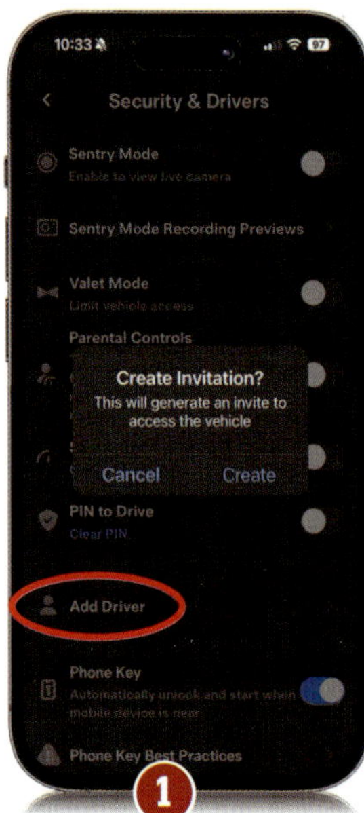

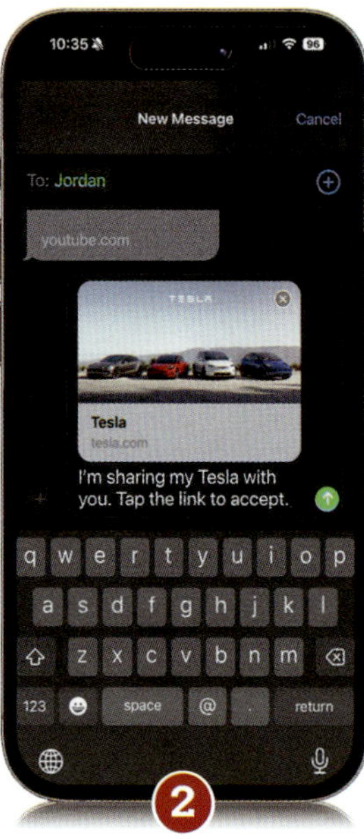

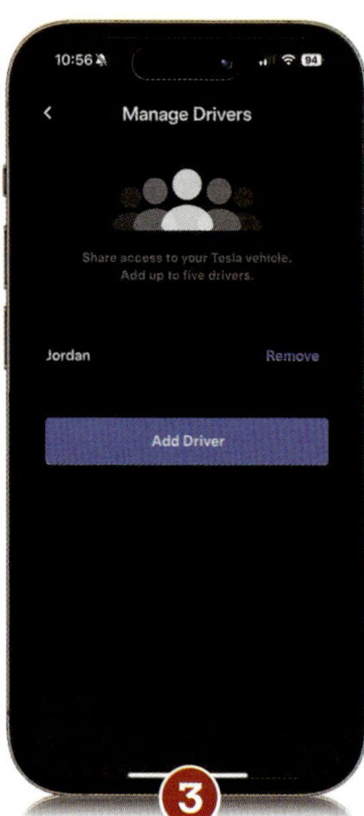

You do this from the Tesla app on your phone. So, launch the app and tap on Security & Drivers, and since this is your first time adding an additional driver, you'll tap on Add Driver (#1 above). This brings up a pop-up asking if you want to "Create Invitation?" so tap the Create button. Once you tap that it gives you a choice of how you want to send your invite (email, text, etc.), and in this case, I chose to text the invite (#2 above). By the way, it writes "I'm sharing my Tesla with you. Tap the link to accept" for you automatically, so all you have to do is add someone in the To field and then hit send. When they accept, and you go back to Security & Drivers in the Tesla app, where it used to say "Add Driver," it now says "Manage Drivers." If you tap on that, it will list anyone you've given driving privileges to (#3 above). To remove a driver, tap on Remove to the right of their name. If you want to add another driver, you'd do that in this screen by tapping the big blue Add Driver button.

How Do I... Make Sure My Driver's Profile Loads Instead of My Spouse's?

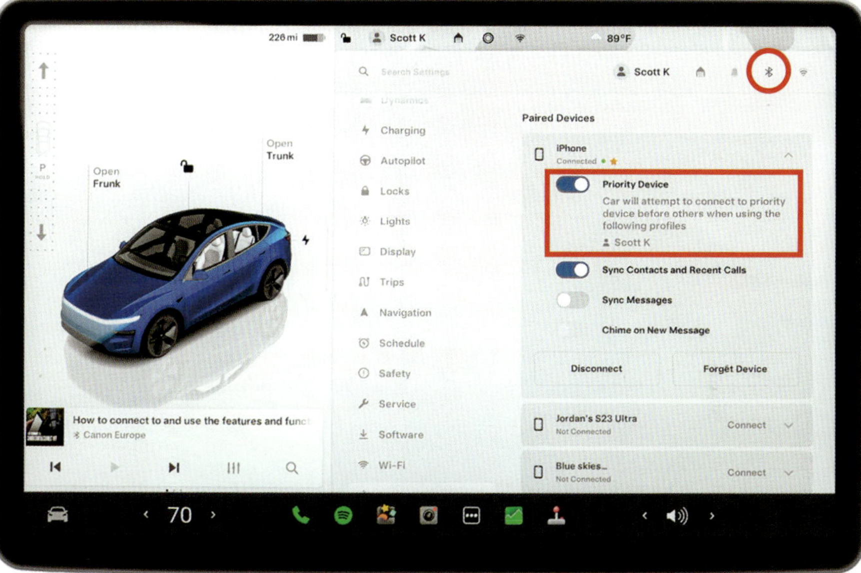

If you and your spouse hop in your Tesla, how does it know which one of you is the primary driver and which driver's profile to load (we looked at creating profiles on page 10)? After all, you both have connected your phones to this car. Actually, it's really easy. If you're the primary driver, you just need to let the car know which phone to connect to first, and to use your driver's profile—including seat adjustments, mirror positions, steering wheel settings, entertainment preferences and so on—rather than your spouse's. To do that, go to your Bluetooth settings by tapping on the Controls icon (the car in the bottom left of your center touchscreen), and then tapping the Bluetooth symbol in the top-right corner. Tap on Paired Devices, then tap on your phone in the list of connected Bluetooth devices, and in the settings, turn on Priority Device. Now it knows to load your profile when you both get in the car.

How Do I... Have My Mirrors Fold In Automatically When I Park My Car?

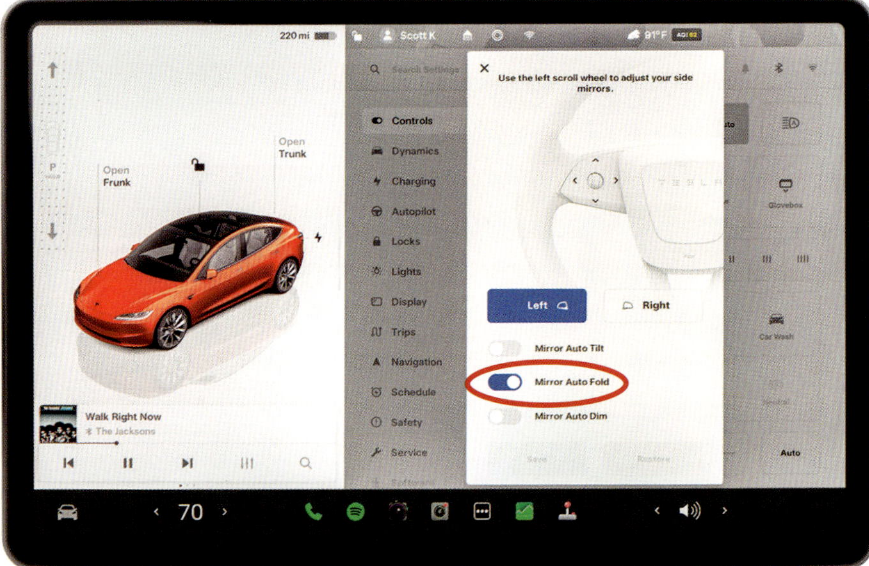

Tap on the Controls icon (the car) in the bottom left of your center touchscreen, then tap on Mirrors, and then turn on Mirror Auto Fold. Now, if you park your car, open the door, and get out, when the doors lock, the mirrors will fold in. When you get back in your car later, the mirrors will automatically unfold. I like having this on because it's a visual way to see that the car is locked when you walk away from your Tesla, since the mirrors only fold in once the car locks.

How Do I... Have My Mirrors Fold In/Unfold Automatically When I'm Home?

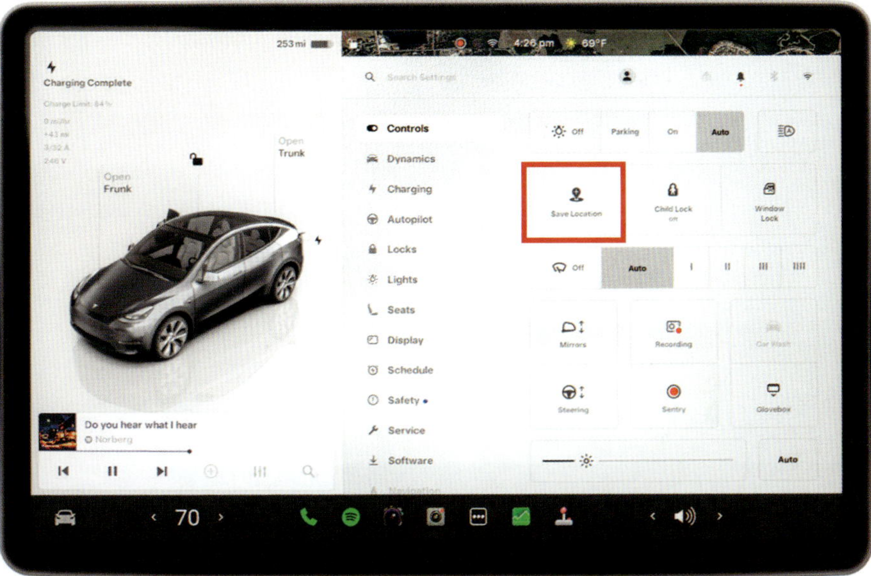

When I come home from work and get close to my garage, my exterior mirrors fold in automatically so I can fit into my spot more easily. When I drive out of my garage in the morning, once I'm out, they automatically unfold. How does it know when to do this? By knowing my car's location based on GPS. It knows when I'm approaching or leaving my garage and it folds them in or unfolds them. Here's how to set this up: First, you can fold/unfold your mirrors anytime by pressing the Microphone button (or the right scroll button in older models) on the right side of your steering wheel, then releasing it and saying, "Fold mirrors" (or "Unfold mirrors"). You can also tap on the Controls icon (the car) in the bottom left of your center touchscreen, and then tap on Fold/Unfold Mirrors (right there on the main Controls screen). Now, how do you get the automatic folding and unfolding based on your location? It's easy. First, drive your car to where you want the mirrors to fold in (for me, that would be in my driveway), then put your car in Park, and then on the Controls screen, tap Fold Mirrors. When you do this, a button appears that says, "Save Location." Tap that and from now on, your mirrors will fold in when you're at that location. It works the same way to unfold them—drive to where you want your mirrors to unfold, tap Unfold Mirrors, then tap Save Location, and you're all set.

How Do I… Open the Trunk from Inside My Tesla?

This will sound silly to say, but I have to say it: this only works when your car is in Park. So, once you're in Park, to open the trunk while you're inside the car, you can either:

(a) On your center touchscreen, where you see the illustration of your car, tap on Open right near the trunk.

(b) Press the Microphone button (or the right scroll button in older models) on the right side of your steering wheel, then release it and say, "Open the trunk."

If you're standing outside your car, you can use the Tesla app on your phone to open the trunk. Tap on Controls and you'll see an overhead view of your car. Tap on the word "Open" that appears over the trunk and it opens. To close the trunk, if you have a newer Model 3 or Model Y, tap the Close button in the inside bottom of the trunk lid. (*Note:* See page 168 for more on using the app.) If you're inside your car, tap Close near the trunk on your touchscreen or just use the voice command "Close the trunk." This is handy if you're dropping someone off and they have something in the trunk or if you're picking up something you ordered online and they bring it to your car and put it in your trunk for you. If you have an older Tesla, you do it old school—you close it yourself by pressing it down until it shuts.

How Do I... Open (and Close) the Frunk (Front Trunk)?

The front trunk (or "frunk," as it's called) has extra storage space (not a ton, but enough for an airport carry-on), and you can open it by either:

(a) On your center touchscreen, where you see the illustration of your car, tap on Open right near the frunk.

(b) Press the Microphone button (or the right scroll button in an older model) on the right side of your steering wheel, then release it and say, "Open the frunk."

If you're standing outside your car, you can use the Tesla app on your phone to open the frunk. Tap on Controls and you'll see an overhead view of your car. Tap on the word "Open" that appears over the frunk and it opens. (*Note:* See page 168 for more on using the app.) To close the frunk, you close it manually by pressing it down all the way, but lightly. Once it's almost closed, then take both hands and press down on it quickly to lock it down. You don't have to press really hard.

How Do I... Have My Trunk (or Frunk) Open for Me?

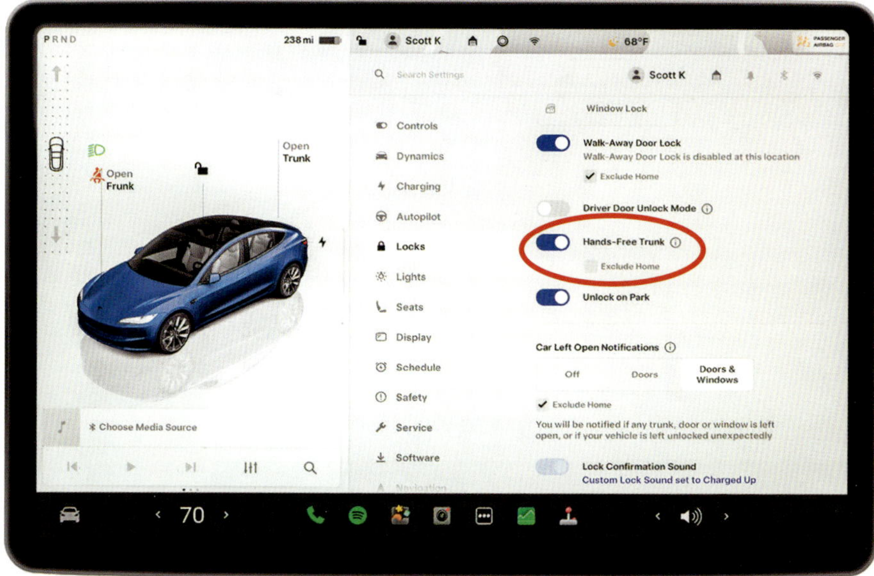

If you have a newer Tesla and an iPhone 11 or newer (as of the writing of this book, Android support hasn't been released yet), you can have your trunk open automatically just by standing behind it (with your iPhone on you) for just a few seconds. (*Note:* You'll need to turn on Nearby Interactions for the Tesla app in your iPhone's Privacy & Security settings.) This is handy if you've got an armful of groceries or packages. To turn this feature on, tap on the Controls icon (the car in the bottom left of your center touchscreen), then tap on Locks, and then turn on Hands-Free Trunk. Once you do that, stand in front of your trunk for a few seconds and you'll hear three short beeps, and then the trunk opens. The frunk will open this way as well (if your model supports it), but it won't pop up the way your trunk will—it will just unlock and pop the hood up a few inches, so your fingers can fit inside and you can lift it manually like always. But, hey—at least it unlocks it without you having to be in your car or launching the Tesla app on your phone.

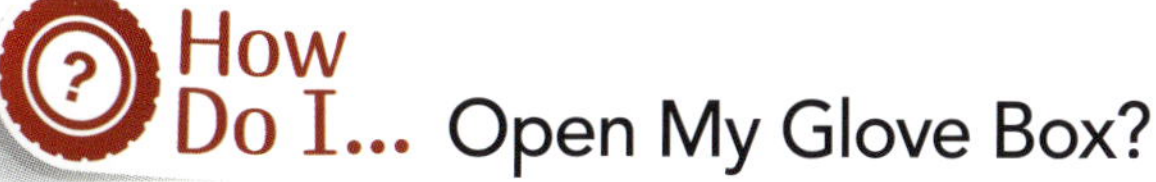

How Do I... Open My Glove Box?

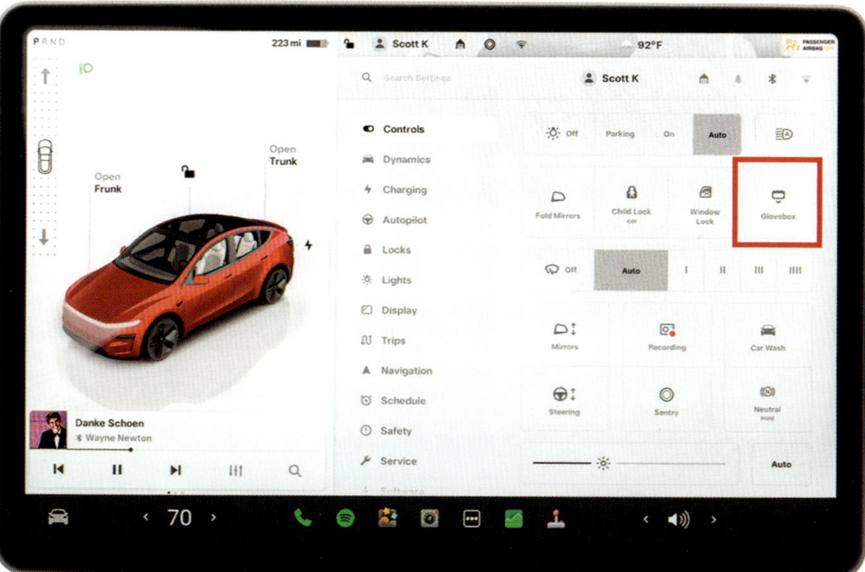

The quickest way (and the way I open it 100% of the time) is to press the Microphone button (or the right scroll button in older models) on the right side of your steering wheel, then release it and just simply say, "Glove box," and it pops right open using that voice command. However, if you want to go old school (well, kinda old school—there is no latch or handle), go to your center touchscreen, tap on Controls (the car icon in the bottom left), and then tap on Glovebox. There is no automatic way to close it—you push it closed manually like you would with any other glove box.

How Do I... Control the Rear Touchscreen from the Front Touchscreen?

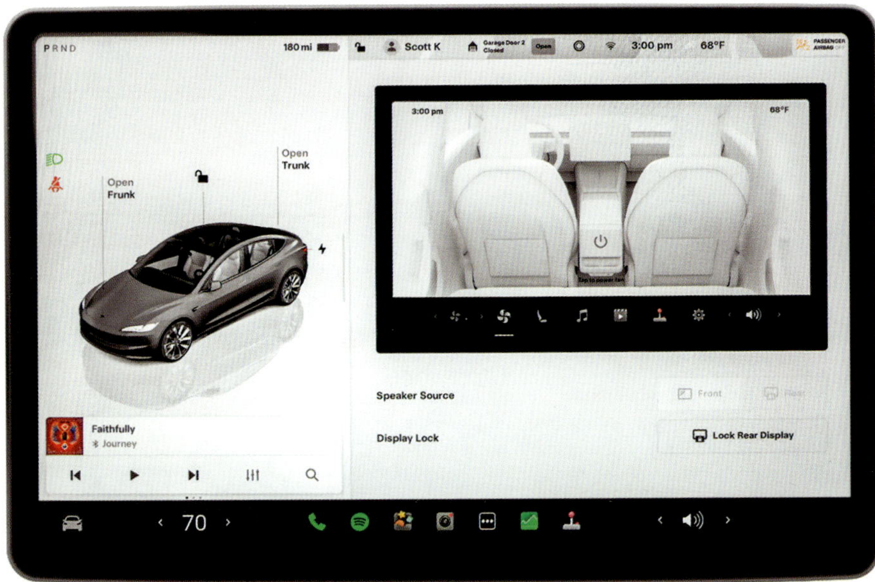

This is super-handy, especially if you have young children in the car, or a guest in the back seat who is not used to working with the rear touchscreen. Tap on the App Launcher (the three dots) icon in the bottom bar of your center touchscreen, and then in the app tray, tap on Rear Screen and this brings up a smaller version of the back seat touchscreen. You have access to all of the controls they have in the back seat, so you can change or adjust anything you'd like from right on your front touchscreen.

TIP: LOCK YOUR REAR TOUCHSCREEN

If your back seat passengers are your kids, and if they start to fight over who is in control of the rear touchscreen or they start messing around with the controls, you can lock that touchscreen from up front on your center touchscreen. Just tap the Controls icon, then tap on Display, and then tap on Lock Rear Display. By the way, you don't have to tell your kids you have this type of control. That way, when you lock it and they say the touchscreen is broken, you can say, "See, now you broke it!" It's stuff like this that makes being a parent totally worth it.

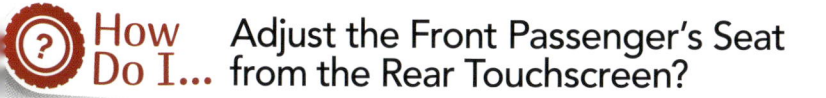

How Do I... Adjust the Front Passenger's Seat from the Rear Touchscreen?

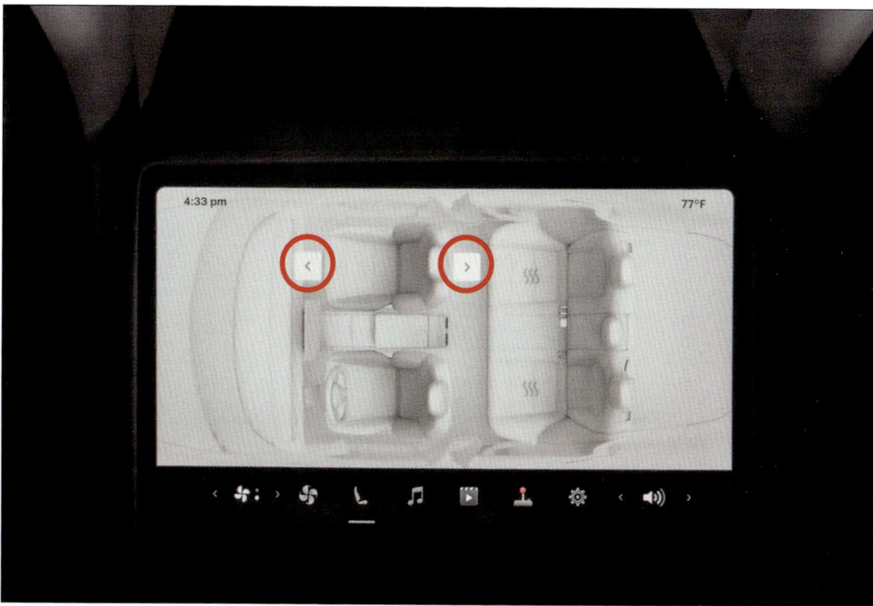

If you have a back seat passenger, but no one in the front passenger's seat (maybe you drive for Uber or Lyft), the back seat passenger can actually move the front passenger's seat forward to create a luxurious amount of leg room for themselves (well, provided they're sitting behind the front passenger's seat). Here's how: on the rear touchscreen, tap on the Seat icon (it looks like a seat), and then tap the arrow in front of the front passenger's seat to move it forward or tap the arrow by its headrest to change the angle of the seat.

How Do I... Fold the Rear Seats?

When you open the back doors and look at the top of the seat closest to you, you'll see a handle. Pull that handle and it releases the seat, so you can simply lay it down until it's flat. However, those back seats have headrests and it's possible you'll have to move either the driver's seat or passenger's seat forward a bit, so those headrests can fully clear those front seats and lay flat. To raise the seats back into position, just grab a seat and pull it back upright. When it's fully upright, it should make a loud click to let you know that it's locked into position, but I always give it tug forward, even after I hear that click sound, to make sure it's 100% locked into position. *Note:* In newer models, you can fold the rear seats electronically by pressing a switch on the left side of trunk, or on the back or corner of a seat.

How Do I... Bring Up the Headrest for the Back Middle Seat (on a Model Y)?

If you have a Model Y and look in the back seat, it might not look like there's a headrest for the middle seat, but there actually is—it's just in a retracted position, by default, so you might not realize there's one there at all. All you have to do to bring it up is grab the top of the middle seat, pull it upward, and it slides on two rails into the headrest position (as seen in the bottom photo here). To put it back down, there's a little button at the base of the first rail (toward the passenger's side). Push that button in, and then press down on the top of the headrest to put it back down in place.

How Do I... Turn On Dog Mode to Keep My Pet Safe While I'm Not in My Tesla?

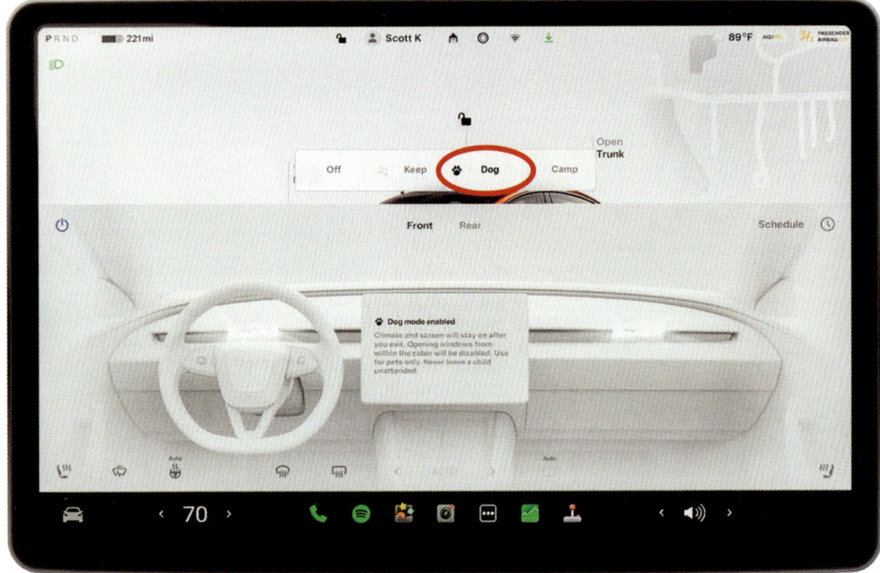

If you're out running errands and need to duck into a store and leave your car for a short period of time with your pet inside, you'll want to turn on Dog Mode. Go to the Climate Controls screen by tapping on either the current temperature or the fan icon at the bottom of your center touchscreen and set a comfortable interior temperature for your pet (cooling or heating, depending on which you need). Then, on the right side of the Climate Controls screen, tap on the Dog button, and then you can exit your car. What's awesome about Dog Mode is that not only does it keep the A/C running so your pet doesn't get overheated, it also puts a large message on your touchscreen to let anyone walking by your car know that the A/C is turned on (it even shows the current temperature in degrees), and that your pet is not in harm's way (it displays in large letters "My owner will be back soon"). It also locks all the windows, so your pet can't accidentally roll them down. (Hey, with a big enough doggo, it could happen, right?) Also, if your battery gets below a 20% charge, your Tesla app will alert you (out of an abundance of caution), so you don't let it get so low that Dog Mode turns off without your knowledge (getting down close to 0% battery). *Note:* Just a reminder, Dog Mode use is designed for short periods of time. Doggos get lonely. Don't leave your doggo (or catto) alone for too long, or they might decide to drive to PetSmart without you.

How Do I... Turn On Camp Mode So I Can "Camp Out" in My Tesla?

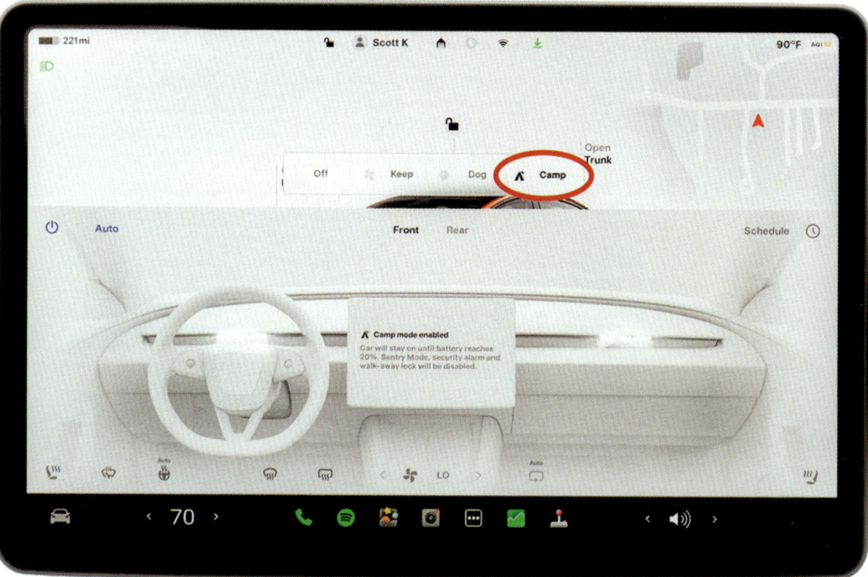

If you want to spend some serious time in your car (anything from watching a movie on your center touchscreen to spending the night camping in it—easier to do if you have a Model Y or X), there's a mode for you: it's called Camp Mode and without having the car actually running, parts of it stay on, like the air conditioning (or heater), the interior lights, the entertainment system, and the USB ports, so you can "go camping" in your car (this is something you absolutely would *not* do in a gas-powered car due to the risk of carbon monoxide poisoning). To turn on Camp Mode, first put your car in Park, and then, as long as you have at least a 20% charge, tap on the current temperature or the fan icon at the bottom of your center touchscreen to bring up the Climate Controls screen. Then, on right side of the screen, tap the Camp button. *Note:* If you're going to be camping out for a few days (people do it), then you'll want to plug your Tesla into a charger so your battery doesn't run out while you're sleeping. If you choose to camp at a campground, they'll often have standard 120v AC power outlets available, and at RV parks, you'll often find higher-powered outlets (just make sure you have an adapter that lets you plug into common AC power plugs—see page 96). Two last things: (1) Camp Mode doesn't automatically lock your doors, so make sure you lock them so that the boogeyman doesn't get you, and (2) for some reason, Sentry mode is disabled when you're in Camp Mode (see page 240 for more on Sentry mode).

How Do I... Lower the Volume of the Interior Car Alert Sounds?

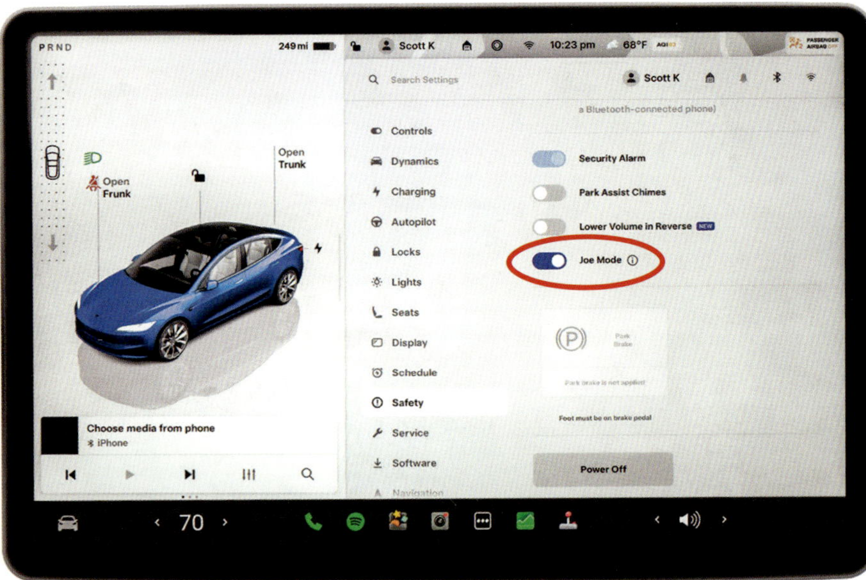

If you need some peace and quiet while driving, there's a special mode called Joe Mode that lowers the volume of many of those things that keep you from enjoying peace and quiet. To enter Joe Mode, tap on the Controls icon (the car in the bottom left of your center touchscreen), then tap on Safety, and then scroll down and turn on Joe Mode. Now, all the chimes and alerts that keep chiming and alerting will still be there, but their volume will be much lower (and thus, more peaceful).

TIP: MUTE THE INTERIOR CAR ALERT SOUNDS

There are lots of little sounds that can vie for your attention while you're driving, and most of them are safety related (like if it detects your car is leaving your lane, or leaving the road, or a car is stopped in front of you and you're approaching it quickly). But, if you want to quickly mute them, go to your center touchscreen and tap the speaker icon that appears to the left of the illustration of your car while it's making those sounds. The icon will change to a speaker with a small "x" next to it, letting you know the alert sounds are muted. To turn them back on, tap that speaker icon again.

How Do I... Customize My Walk-Away Lock Sound?

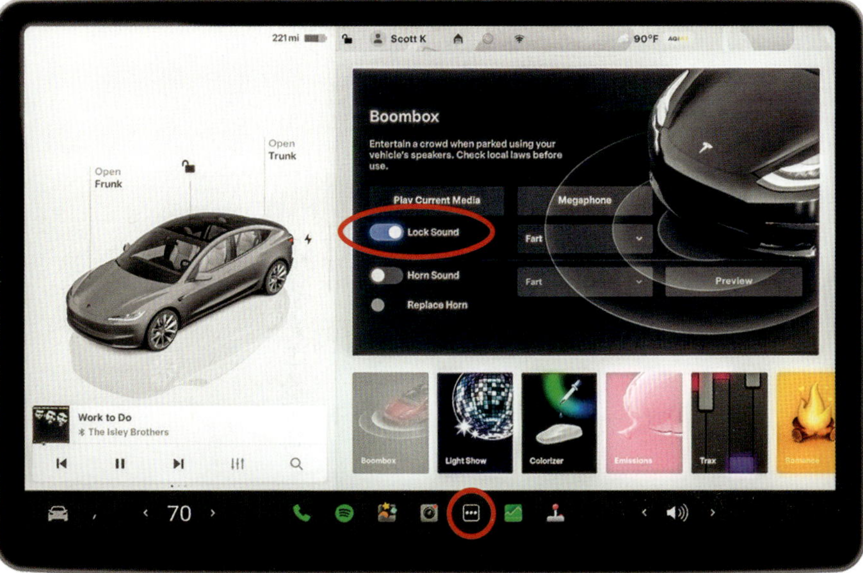

When you walk away from your Tesla, it automatically locks the doors for you, and the lights flash once to let you know that it did, indeed, lock (of course, you have to turn around and look to see that this happened). However, if you'd feel better hearing an audible sound that lets you know that, "Yes, your car is locked," then you can turn this feature on by tapping on the Controls icon (the car) in the bottom left of your center touchscreen, tapping Locks, and then turning on Lock Confirmation Sound. With this on, you'll hear a single "beep" to let you know things are locked up safe and sound. But hold on folks, there's more: You can change the sound it makes when it locks by tapping on the App Launcher (the three dots) icon at the bottom of your touchscreen, then in the app tray, tapping Toybox, and then tapping Boombox. On this screen, turn on Lock Sound and a pop-up menu of built-in sound choices appears (seen above). Choose one you like, and now when you walk away and your car locks, it will play this custom sound (instead of beeping the horn) through your Tesla's external speaker (most Teslas have one). If you want to take things up a notch, and you're a tad of a computer nerd, you can download other sounds from sites like https://teslapro.hu/lockchimes/ and save them as AVI files through the USB flash drive that's in your glove box. (*Tip:* Save them without the ".avi" as the filename.) By the way, your doors automatically lock when you're driving—as soon as your car goes over 5 mph (I guess they felt they should wait and leave them unlocked when you're going under 5 mph, so you can open the door, jump out of the moving car, and roll onto the ground, like you see on detective shows).

How Do I... Unlock My Doors While I'm Inside My Car?

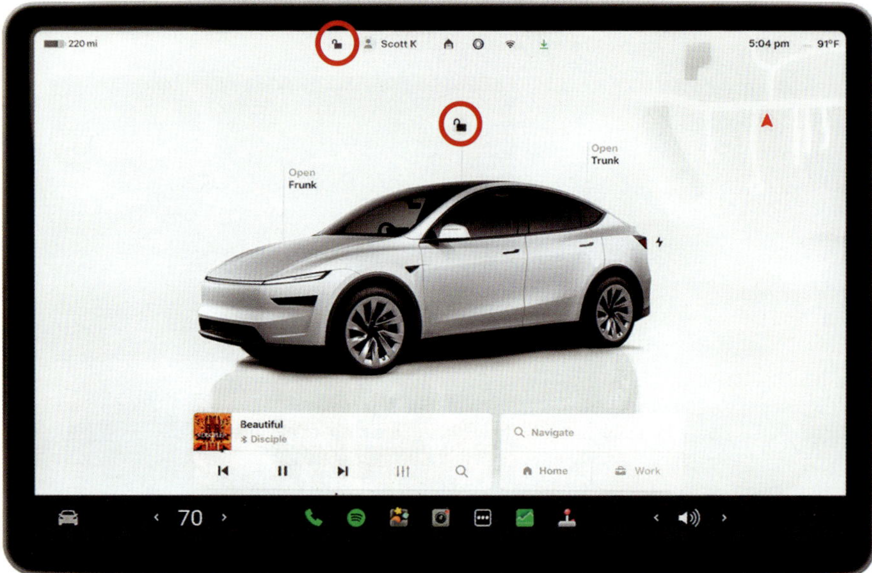

Just tap the Lock icon near the top of your center touchscreen and the doors will unlock. Of course, you can also do this from the Tesla app, as well (tap the Unlock icon that appears below and to the left of the illustration of your car). Lastly, you can use a voice command by pressing the Microphone button (or the right scroll button in an older model) on the right side of your steering wheel, then releasing it and saying "Unlock the doors."

TIP: UNLOCKING ALL THE DOORS WHILE YOU'RE IN PARK

While you're in Park, you can also tap on Park again on your center touchscreen to unlock all the doors.

How Do I... Get a More Comfortable Ride Overall?

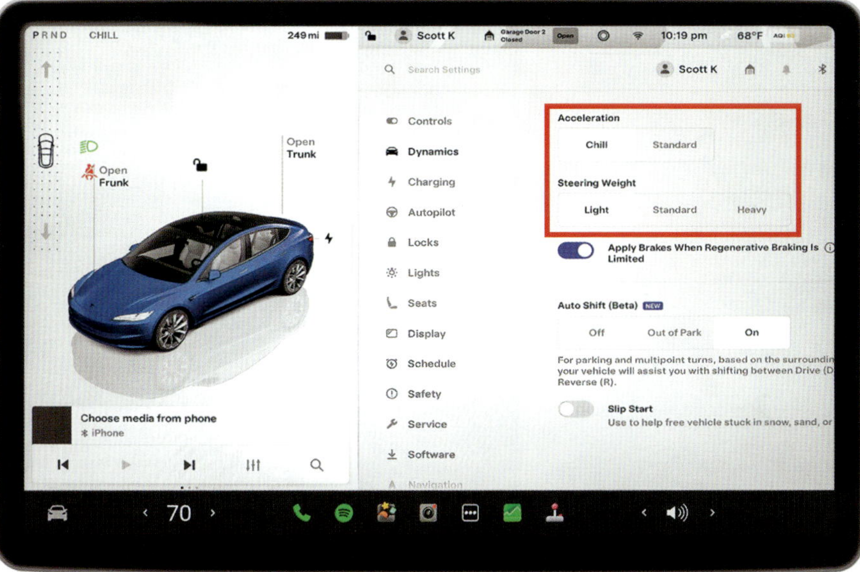

If you want a smoother, more comfortable experience when you drive, there are two settings you can change to get you there: The first is called "Chill" acceleration, which essentially takes the "bite" out of the acceleration when you press the gas pedal. By default, your car is in Standard acceleration, and when you step on the gas, that baby takes off, but Chill is a more relaxed version of that. The second setting to change is the steering weight. You'll switch it from Standard (or Medium) to Light, which gives you a smoother, easier, and somewhat more luxurious feel to steering by reducing how much effort it takes to turn the wheel. You'll notice this easier turning most when driving at slow speeds or parking your Tesla. You get to these settings by tapping on the Controls icon (the car) in the bottom left of your center touchscreen, then tapping on Dynamics (or Pedals & Steering, depending on your model). There you'll see the settings for Acceleration (choose Chill) and Steering Weight (choose Light).

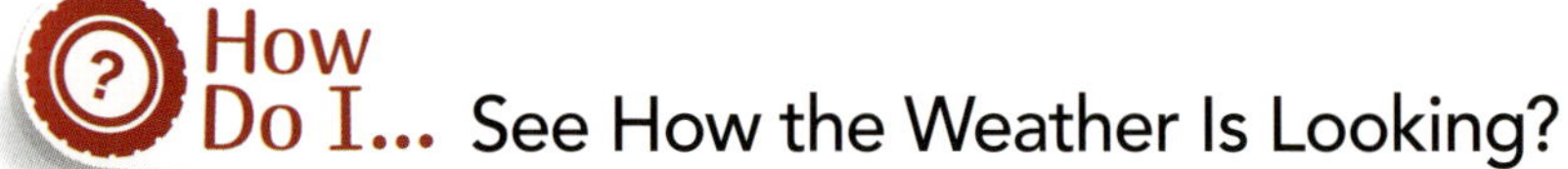

How Do I... See How the Weather Is Looking?

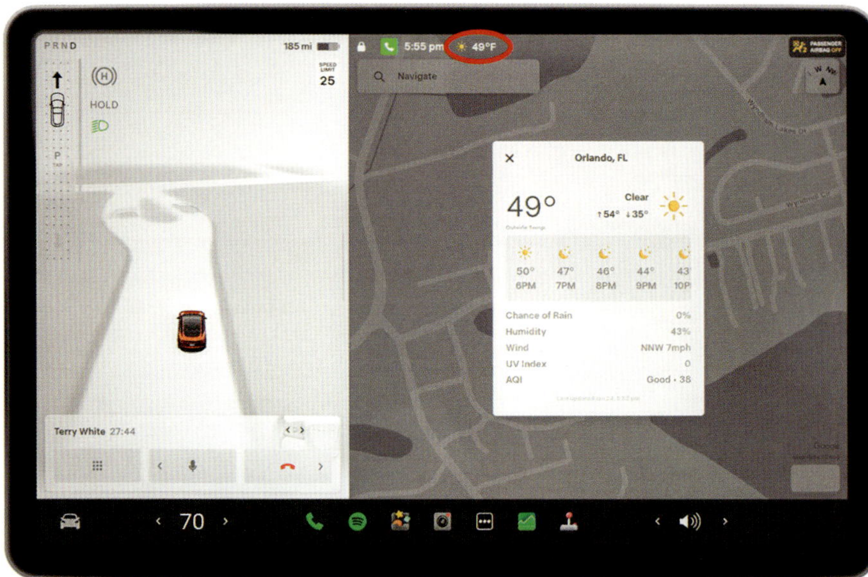

If you're curious about what the weather is like outside, you can either (1) look in the row of icons in the Status Bar at the top of your center touchscreen, where you'll see the temperature and an icon that reflects the current weather—so you might see a sun icon, or a rain cloud icon, or a snow icon, etc. The other method (2) is to just look out your window and you'll see what the weather's like where you are. (Oh, come on, that one was funny.) If you want more detailed weather information, just tap on that weather icon, and it brings up a window with the high and low temps for the day, the upcoming forecast for the next few hours, the rain chance, the wind speed and direction, and more. When you're done, just tap on the "X" in the top-left corner of the window to close it. (*Note:* See page 153 for how you can see a live radar weather map.)

How Do I… Turn On My Overhead Dome Lights?

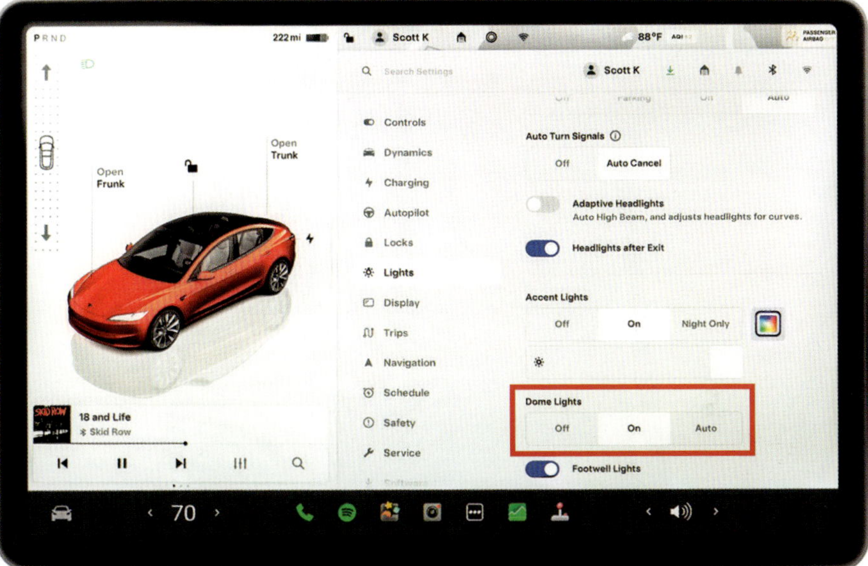

By default, they're set to Auto, so they automatically come on when you open any of the car doors or put your Tesla in Park, but if you want to turn them on at any time, you can:

(a) Use a voice command by pressing the Microphone button (or the right scroll button in older models) on the right side of your steering wheel, then releasing it and saying "Dome lights."

(b) Go to your center touchscreen, tap the Controls icon (the car in the bottom left), then Lights, and then beneath Dome Lights, tap On.

(c) Simply tap directly on one of the dome lights themselves (yup, you just reach up there and tap on one) and it'll come on.

How Do I... Find My Cigarette Lighter (the 12V Power Socket)?

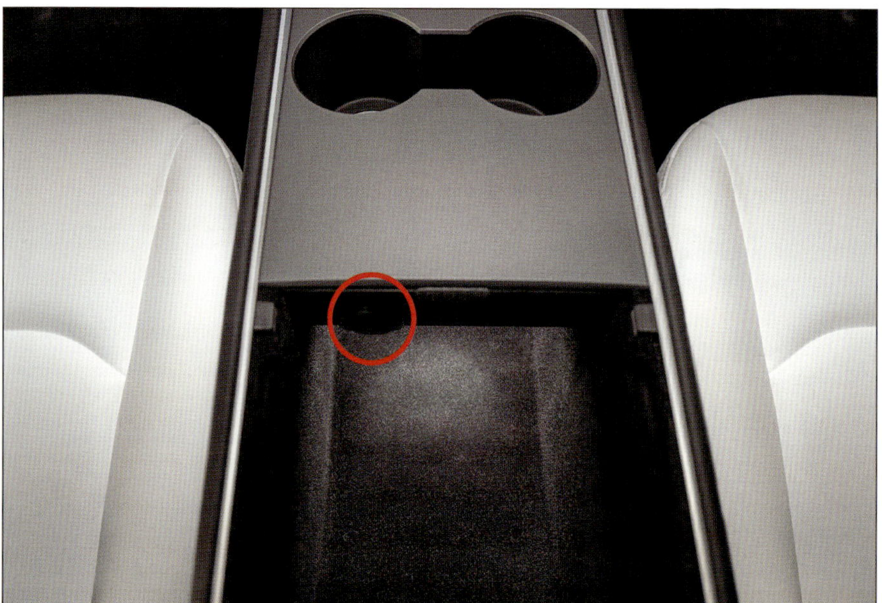

If you need to plug in something that uses an old-school cigarette lighter (for example, I travel with a little device that pumps air into a low or flat tire that requires this type of plug—see page 268), the 12V power socket is found in the front console and it's in the section where you lift up the armrest. Look at the little flat wall separating this compartment from the other compartments and you'll see that 12V cigarette lighter/power socket.

How Do I... Charge My Phone?

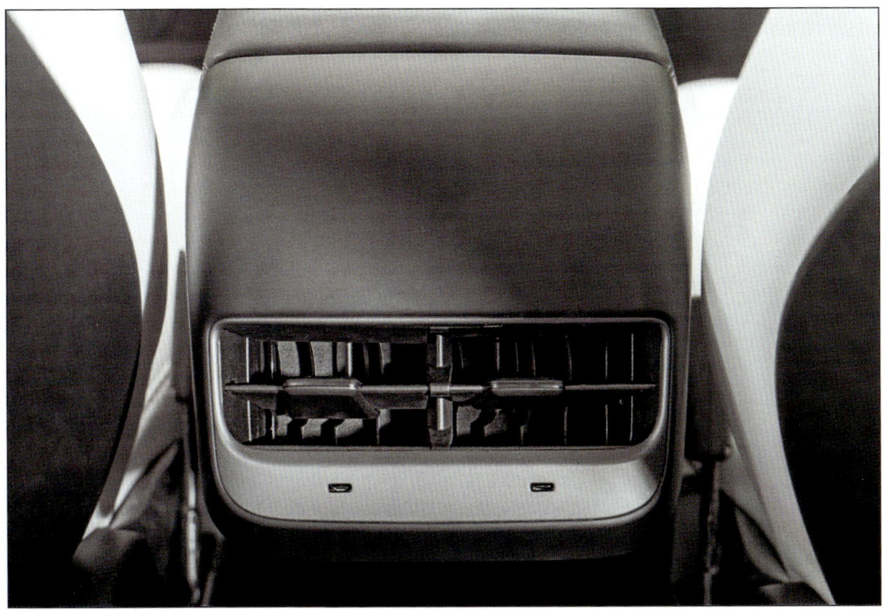

There are a number of places to charge your phone in a Tesla (apparently, they really want your phone to be charged). If your phone accepts wireless charging, you can simply place it on one of the two wireless charging pads at the front of the console between the two front seats. A small indicator light comes on near the bottom center of the charging pad to let you know that it's charging. (*Note:* If you have a thick case on your phone, it might not work with this wireless charging.) While these are very convenient, they are the slowest charging slots in the car (they only charge at 15W). There is also a much faster USB-C charge port(s) inside the front console (you can see it when you slide back the little sliding door—the port is facing you in the center. You may have one or two, depending your model). You'll plug a USB-C charging cable into that slot and connect your phone to the other end of the cable. There's an old-school USB-A port inside the glove box (on the back left side of it), plus there are two USB-C ports for your rear seat passengers beneath the back seat air conditioning vents.

How Do I... Make a Phone Call from My Car?

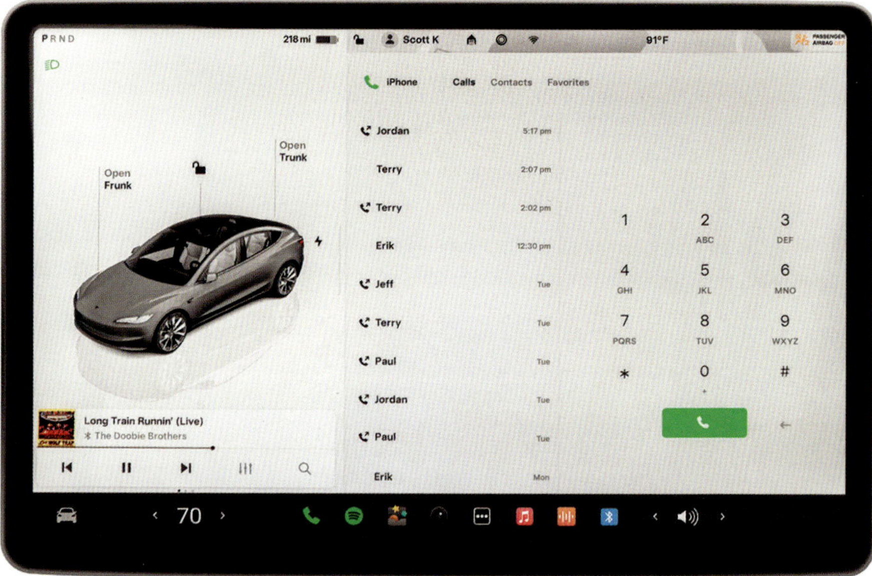

If your phone is connected to your car via Bluetooth (and it should be, right?), then this is super-easy (if it's not, see page 35). Press the Microphone button (or the right scroll button in older models) on the right side of your steering wheel, then release it and say, "Call Scott" and it starts dialing. Yes, it's just that easy. But, I'll tell you right now—I'm probably not going to answer, so you should call somebody else who is more likely to answer (like your mom or my mom). Anyway, once they answer, it's all hands-free and you'll hear them through your Tesla's audio system, and you'll answer back via the built-in mic. You control the call volume using the left scroll button on your steering wheel. You'll also see a card (mini-panel) appear in the bottom left of your center touchscreen with their name, and controls for hanging up or muting your mic. If the person you want to call isn't in your address book (you did choose to Sync Contacts and Recent Calls in the Bluetooth settings panel, right? If not, tap the Bluetooth icon at the top of your touchscreen, then in the window that appears, tap on your phone in the list, and then turn that feature on), then you can just type in their number manually. To do that, tap on the App Launcher (the three dots) icon in the center of the bottom bar of your touchscreen, and in the app tray that pops up, tap on Phone. This brings up a keypad like your phone has where you can simply type in the number you want, and then tap the call button at the bottom. There are also tabs up top to show your contacts, see recent calls, and see your calling favorites. One more thing: if someone calls you while you're driving, you can answer by tapping the phone icon that pops up in a card in the bottom left of your touchscreen.

How Do I… Have My Fan Speed Automatically Turn Down When I'm on a Call?

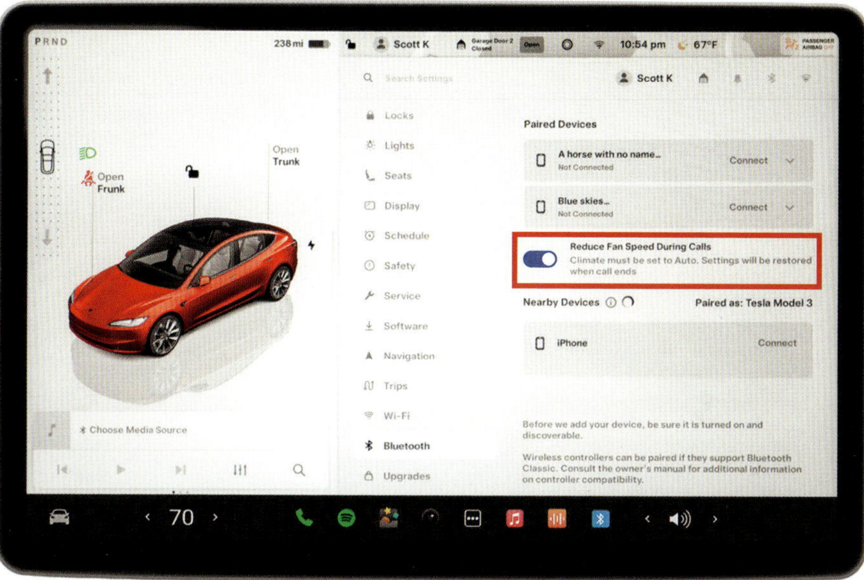

If you have your Climate Control set to Auto (where it automatically cools or heats your Tesla to a temperature you've chosen; see Chapter 3 for more on climate), the fan can be pretty loud when you first get in your car, as it's doing its most intense cooling (or heating) since your car has just been sitting there. If you want to make a phone call, all that noise can make it hard to hear, but your Tesla can lower the fan speed automatically any time you're on a call. To turn this feature on, tap on the Controls icon (the car in the bottom left of your center touchscreen), then tap on Bluetooth, and then turn on Reduce Fan Speed During Calls. When you're finished with your call and hang up, the fan will go back to its previous speed.

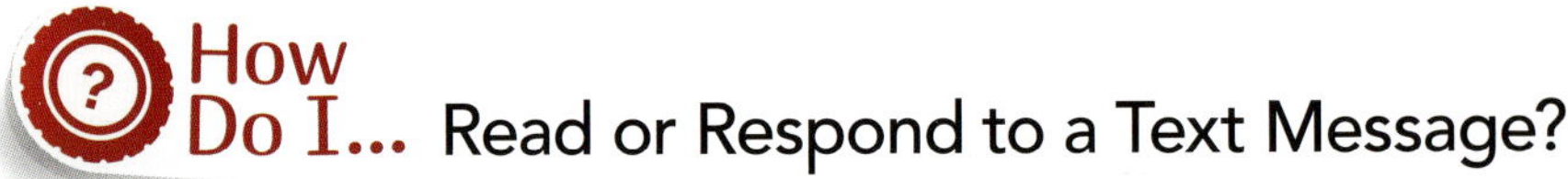

How Do I... Read or Respond to a Text Message?

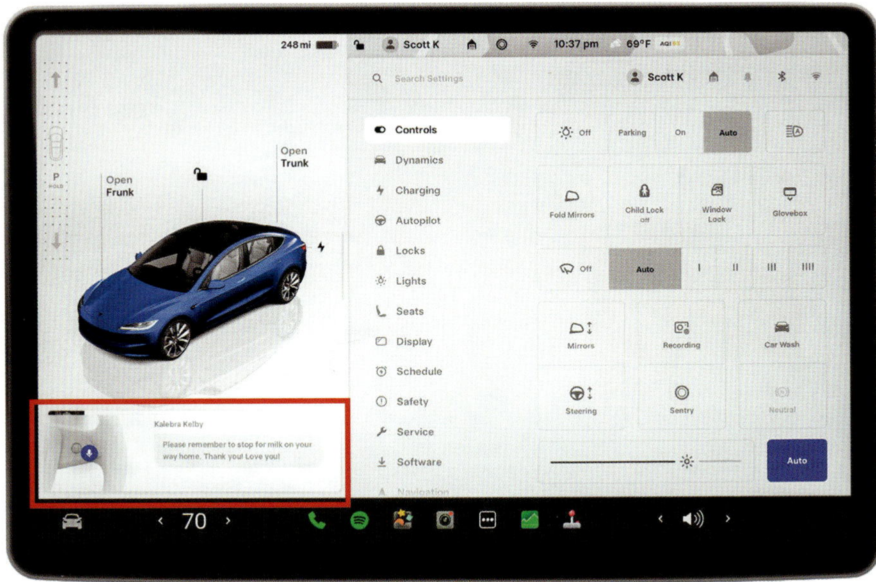

I probably don't even have to mention this, but the first step is to make sure your phone is connected to your Tesla via Bluetooth, and that in your Tesla's Bluetooth settings (in the Controls screen), you have your Sync Contacts and Sync Messages switches toggled to On (I know you knew that, but I had to say it. See page 35 for more on this). You'll also need to have Notifications turned on in your phone. To send a text message, press the Microphone button (or the right scroll button in older models) on the right side of your steering wheel, then release it and say, "Send a text message to...," followed by the name of the person you want to text (they need to be in your phone's address book). Then, simply say what you want to text, and you'll see your message appear in the bottom left of your center touch-screen as you talk (it will look like a regular text message). If you make a mistake (or say something you'll regret later), just double-press the Microphone button to erase what you said and start speaking again. When you're done, press the Microphone button once to send the message. To have your Tesla read you an incoming text message, first go back to your Bluetooth settings, and make sure Chime on New Message is turned on, so it plays an audible chime sound when a new text comes in. To hear the message, press the Microphone button once. To skip it, press that button twice. To reply to that message, press the Microphone button once, and then just say your reply, which you'll see in the bottom left of your touchscreen, and then press it once more to send your message. You can also see your messages on your touchscreen by tapping on the App Launcher (the three dots) icon at the bottom of your touchscreen, and then in the app tray, tapping the Messages app.

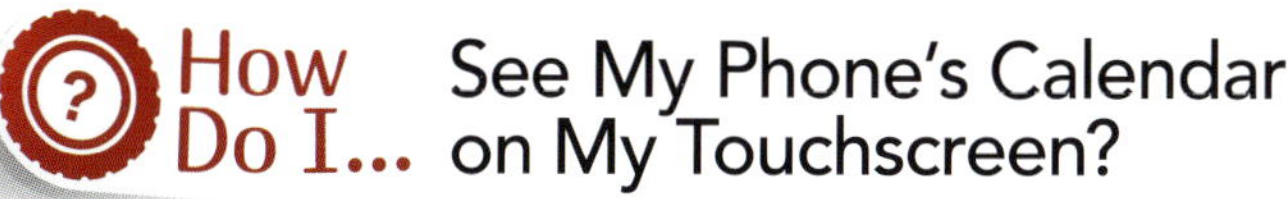

How Do I... See My Phone's Calendar on My Touchscreen?

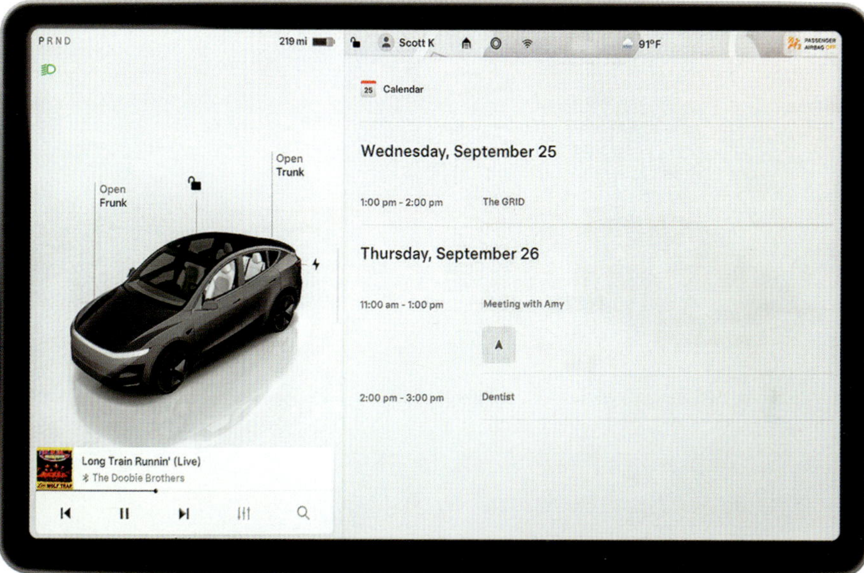

When you connected your phone to your Tesla via Bluetooth (see page 35), one of the options was to sync your calendar, which I love because when I get into my car in the morning, I can take a quick peek at what's on my calendar for that day. I especially love it if that day is blank (clearly, it wasn't today, or this page would be blank). But before we go digging through menus, start by going to the Tesla app on your phone, tap the hamburger icon (that's the three lines in the top-right corner) to bring up a menu and at the top is your account. Tap on that and at the very bottom of your account screen, make sure Calendar Sync is turned on (if it's not, well...turn it on). Okay, now, back to the car. To see your calendar, just tap on the current time at the top of your touchscreen, and your calendar appears onscreen. That's all there is to it. Here's wishing you lots of blank calendar days (or days that say "Off on vacation").

How Do I... Find Features/Controls When I Don't Know/Remember Where They Are?

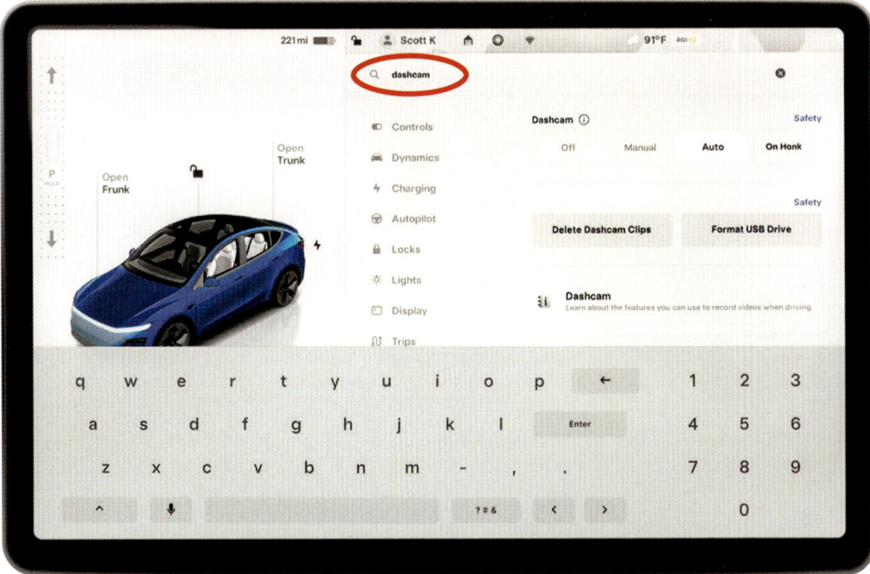

Your Tesla has a lot of features and a lot of menus, and that's why you'll love to know that when you tap on the Controls icon (the car) in the bottom left of your center touchscreen, at the top is a Search Settings field. Tap once in that search field to bring up the onscreen keyboard, type in what you're looking for, and it brings you that feature so you can use it right there, or you can tap its blue link to take you to the screen where it lives.

How Do I... Name My Tesla?

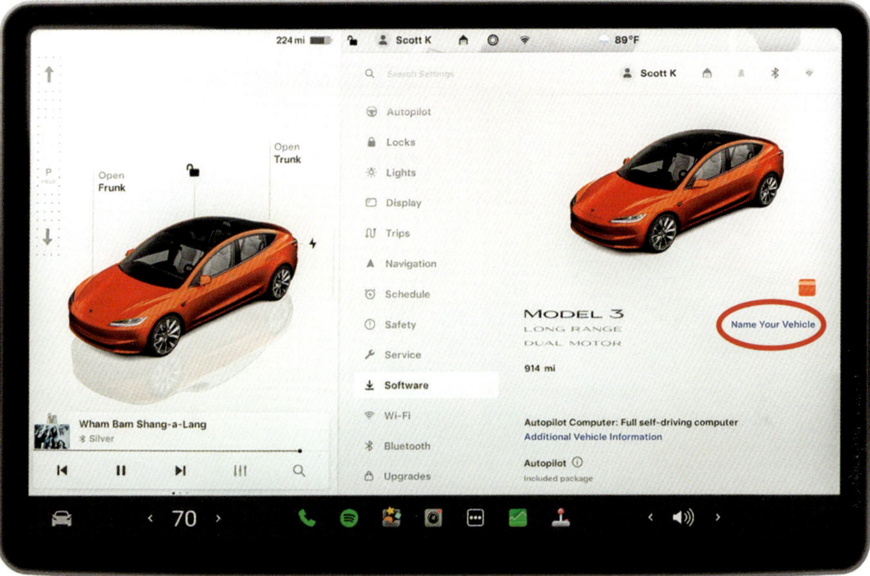

If you want to give your Tesla a name (rather then its generic factory-given name), you do that by tapping on the Controls icon (the car in the bottom left of your center touchscreen), and then tapping on Software. To the right of your car's model on this screen, over on the far right, you'll see some blue text that says "Name Your Vehicle." Tap on that and a naming window pops up. Tap in the naming field and a keyboard appears where you can type in the name you prefer. When you're done, tap Save and your car is named (and you'll see that reflected in a number of places, including in the Tesla app). That's all there is to it.

Change My Tesla's Illustration Color on My Touchscreen?

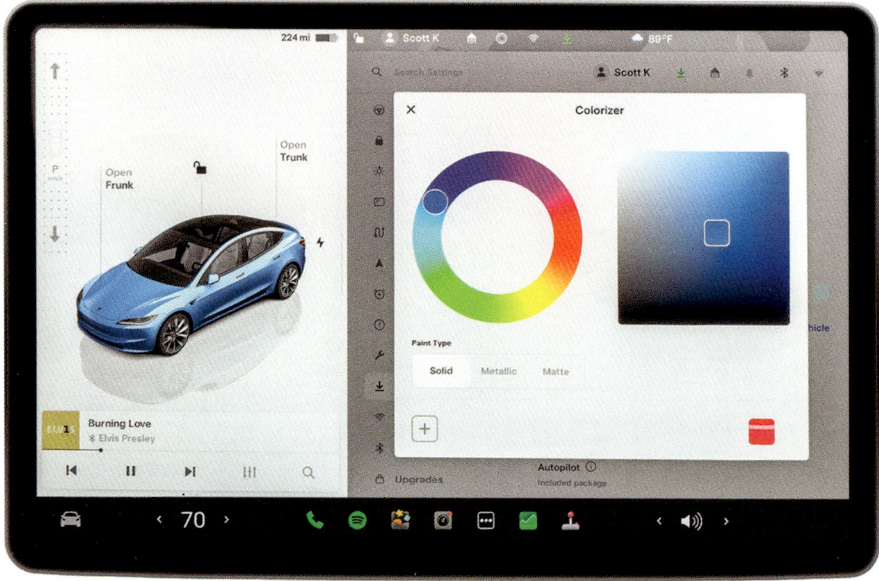

By default, the 3D rendering of your Tesla that you see on your center touchscreen is based on your actual car—so the same model, the same wheels, and of course, the same color. If you'd like to change the color of your rendering, tap on the Controls icon (the car in the bottom left of your center touchscreen), and then tap on Software. Right above the words "Name Your Vehicle" (or whatever you named your Tesla; see the previous page) on the far right, you'll see a small, square color swatch (if you have a white car, it blends in with the background to where you'd hardly realize it's there). Tap on that color swatch and it brings up the car Colorizer (color picker), where you can choose any color you'd like (including colors that aren't available from Tesla in real life). The ring on the left side chooses the hue and the square lets you choose the saturation for the color. There are also tabs for Metallic colors or even Matte colors. When you're done "colorizing," tap the "X" in the top-left corner to close the window and apply your new color. By the way, an "unauthorized color change" wouldn't be a bad prank to pull on one of your Tesla-owning friends.

TIP: SEE MORE OF YOUR CAR IN 3D

That 3D rendering of your Tesla is really in 3D, and you can pinch to zoom in on it, and even rotate it around to see different views and angles of your car by holding your finger right on the car and dragging left, right, up, and down.

How Do I... Control the Wraparound Accent Lighting?

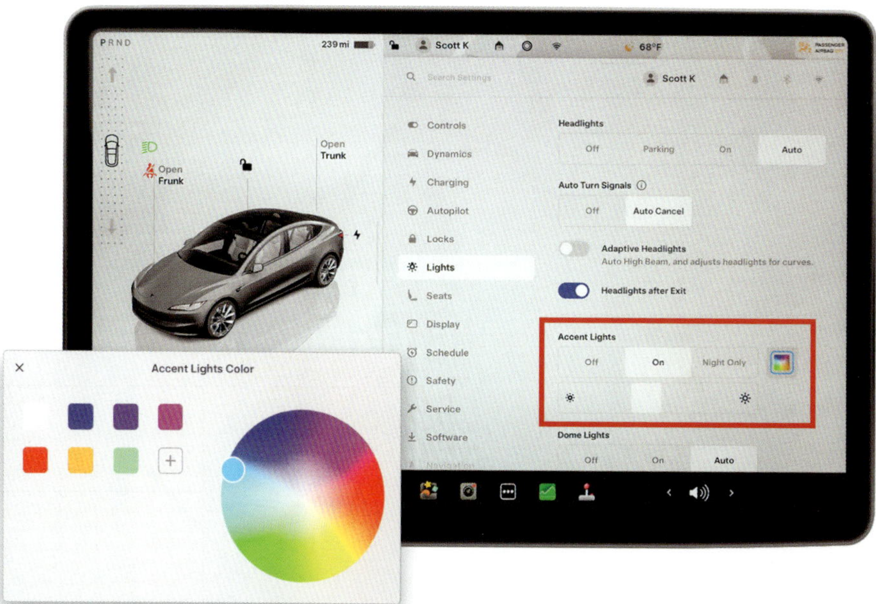

If you have a newer Model 3 or Model Y, you have color accent lights that wrap-around the interior of your car, and not only can you control how bright these accent colors are, you get to choose any color you'd like them to be. You do this by tapping on the Controls icon (the car in the bottom left of your center touch-screen), then tapping on Lights, where you'll find Accent Lights. You can turn them Off (Booo! Booo!), turn them On, so they're visible all the time, or you can set them to Night Only. Below the buttons is a slider that you can tap-and-drag to control the brightness and to the right of those buttons is a color wheel icon. Tap on that icon to bring up the Accent Lights Color picker you see in the inset above. Just tap-and-drag the small circle on the big color wheel to the color you want, and the accent lighting responds instantly so you can see how that color looks. You can also save any color as a favorite by tapping the + (plus sign), and it will save the current color as one of those color tiles to make choosing that exact same color just a one-tap proposition next time.

Chapter 3

How to Use Climate Control Like a Boss

Too Hot? Too Cold? Too Bad!

I know I used the phrase "Like a boss" in the title here, and although I still use that phrase today, don't go by me because I also still use phrases like "Whoomp! There it is" and "Who let the dogs out?" along with "Better safe than sorry," "Whoops-a-daisy," and "What you talkin' 'bout Willis?" So clearly, I am "hip with the kids." It just didn't seem to me that "Like a boss" was that old, but I just looked it up and that phrase started trending on Twitter (now X) back in 2009. Ouch! Do you know which car Tesla made back in 2009? It was the original Tesla Roadster, which was a Tesla powertrain and battery with a Lotus body on top. But, what surprises folks is that, back then, they didn't have the elaborate battery system modern-day Teslas have. Instead, that car was powered by a single 9-volt battery—the kind you put in a transistor radio! What's a transistor radio, you might ask? It's a small, hand-held radio old folks used back in two thousand aught nine to listen to AM radio. What's AM radio, you might ask? It's a radio transmission that only broadcasts "awful music" (which is where the "AM" moniker came from). It was mostly static and noise, but if you listened intently, you could occasionally pick out a musical note. It was a brutal time. That's why people mostly listened to talk radio on AM, and one talk show I loved was called *Cooking in Your Tesla*, about two topics I'm passionate about: cooking (when somebody else does it) and Teslas. It worked because, back then, the battery got hot enough to fry an egg. Anyway, the recipes they shared were yummy, but somewhat dangerous due, in part, to how careful you have to be when cooking inside a moving vehicle. One small mistake during the process could leave you with a real mess on your hands, which turned off many listeners. Anyway, if you did cook up something tasty in your Roadster, you would never forget the feeling you had when you'd take it out of your car, place it in the center of your dining room table, and proudly exclaim, "Like a boss!" and all your friends and family would cheer because that was such a cool phrase back then. Those were good times, my friend. Good times.

How Do I... Adjust the Air Conditioning?

Tap the temperature in the bottom left of your center touchscreen and on the Climate Control screen, you'll see the current direction of the airflow from the horizontal, face-level vent that spans the width of your dashboard. To readjust where it's blowing, just tap-and-hold directly on the vent and drag it where you want it. You can also pinch in for a more direct airflow or spread your fingers outward for a wider spread of air (this option alone made this my favorite air conditioning feature ever!). At the top of this screen are icons that you can use to choose if you want the airflow to come from the front vents, only from the floor vents, or from the windshield vents (and this isn't an either/or; you can make multiple choices here). You can adjust the intensity of the fan by tapping on either arrow on the sides of its icon (or using the pop-up sliders that appear when you tap on one of those arrows). Tap on Auto to have it adjust the fan amount to the temperature you have set, and you can set the driver's side and passenger's side temperatures individually by tapping on the temperature, and then tapping on Split in the pop-up (more on this on page 80).

TIP: QUICKLY TURN OFF CLIMATE CONTROL

Just tap-and-hold the temperature reading in the bottom left of your center touchscreen for three seconds.

How Do I... Turn On the Air Conditioning for My Rear Passengers?

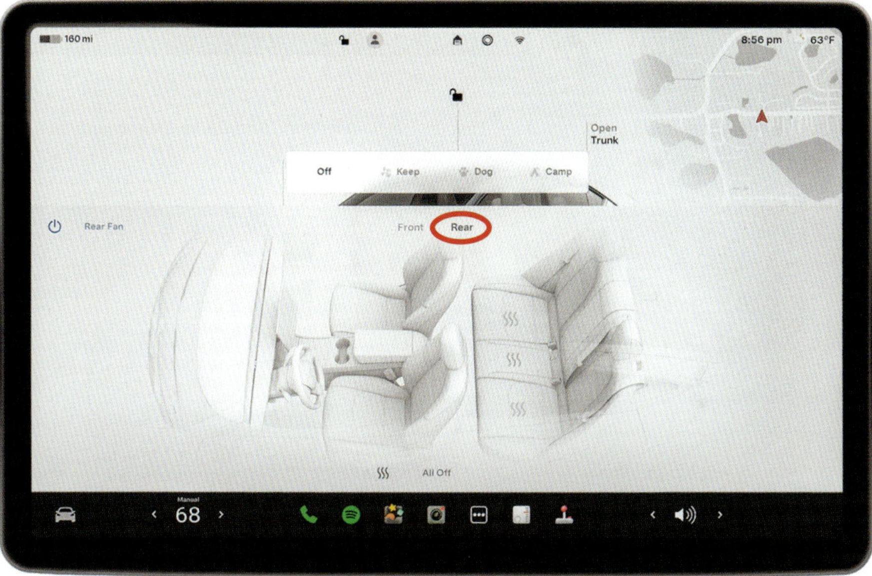

In the bottom left of your center touchscreen, tap on the temperature reading to bring up the Climate Control screen and you'll see a Rear button that's grayed out—that's the button that turns on the rear seat climate control. So, just tap on that to turn it on to start the airflow to the back seat passengers. The back seat vent moves air at the same fan setting you have for the front seat passengers, so if it gets too cold back there, you'll have to lower the fan setting for the entire car from your touchscreen in the front seat. If you have a newer Tesla (and depending on your model), see page 85 for how to let the backseat passengers control their own climate settings.

How Do I... Schedule My A/C to Automatically Turn On Before I Get In?

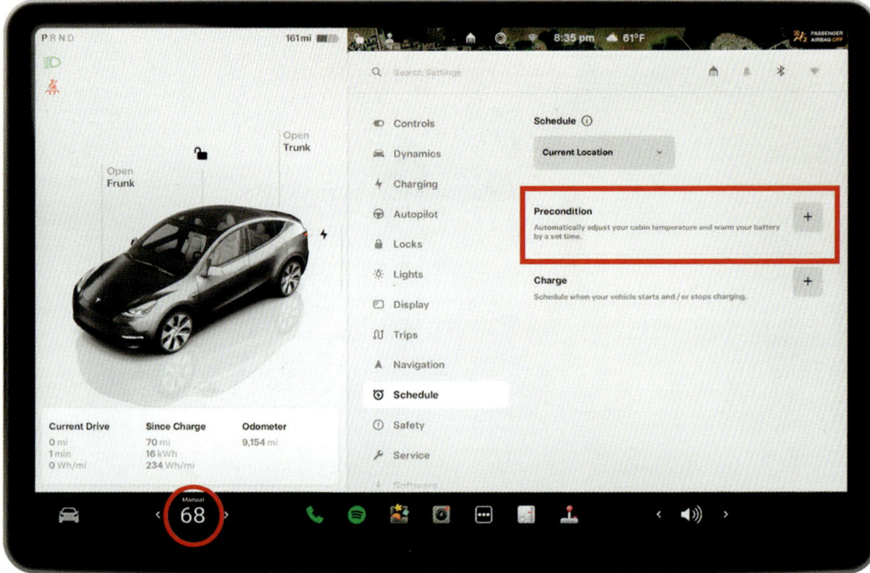

Tap on the temperature reading in the bottom left of your center touchscreen, and then in the Climate Control screen, tap on Schedule near the top right. This brings up a screen where, under Precondition, you can choose what time you'd like to have the air conditioning (or heat) come on automatically, so your Tesla is cooled (or heated) to the temperature you choose before you ever get in. (*Note:* See page 172 for how to do this from the Tesla app.)

How Do I... Keep My Tesla's Temperature Cool (or Warm) While I Dash into a Store?

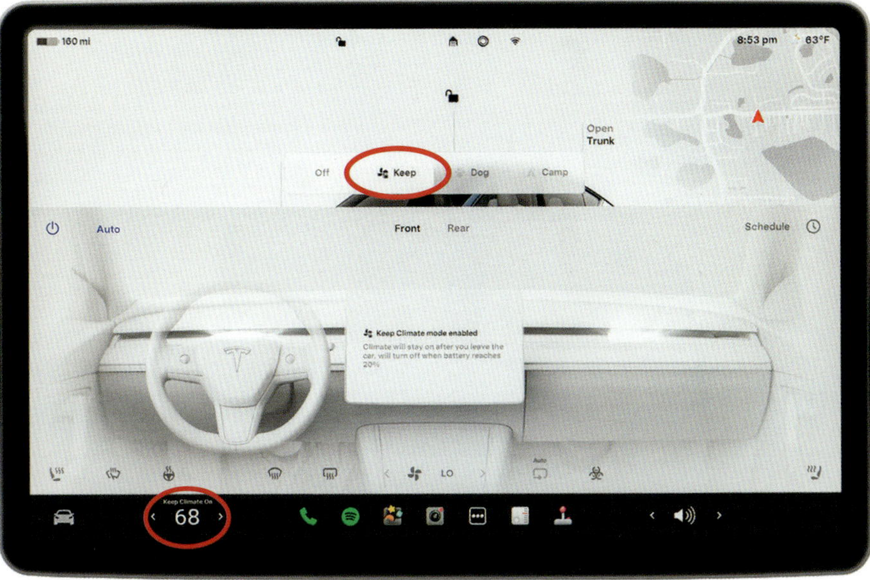

I live in Florida, so this one is a lifesaver during the summer because I can leave the A/C on to keep my Tesla cool while I run into Dunkin' for a coffee and a circular-shaped accompaniment. To do this, tap on the temperature reading in the bottom left of your center touchscreen to bring up the Climate Control screen, and then tap on Keep, which tells your car to keep the A/C running even when you leave it. You'll know it's turned on when you look at the temperature setting—you'll see the words "Keep Climate On" appear right above the temperature (as seen here). Also, although running the A/C uses battery power (as does just about everything), don't worry, you won't get stuck. If your battery gets below a 20% charge, it will turn off this feature automatically to conserve battery.

How Do I... Set the Driver's and Passenger's Temperatures Separately?

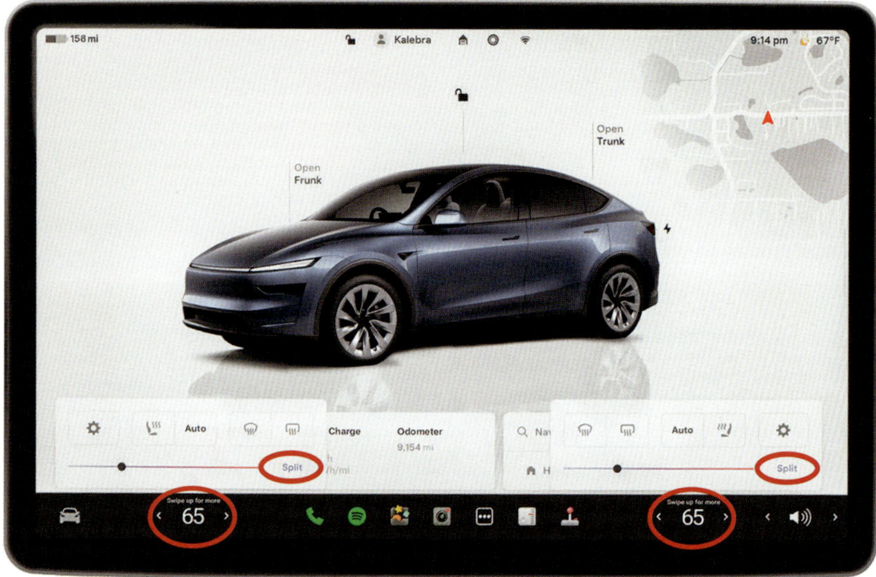

In the bar along the bottom of your center touchscreen, you'll see temperature readings on both ends (the one on the left shows the driver's side temperature setting and the one on the right shows the passenger's) when you have Split turned on. If you raise the driver's side temperature, by default, it's in sync with the passenger's side temp (for example, if it's set at 71° and you move the driver's side to 73°, the passenger's side temp moves to 73° as well). To split them, so each of you can choose your own temperature for your side of the car, tap once on either arrow icon on the sides of the temperature, and then in the Climate Control pop-up, tap on Split. Now the two temperature settings you see in the bar at the bottom of your touchscreen are no longer in sync, so your passenger can use the arrows next to the temperature on their side to adjust it on their side of the car without any effect on the driver's side temperature.

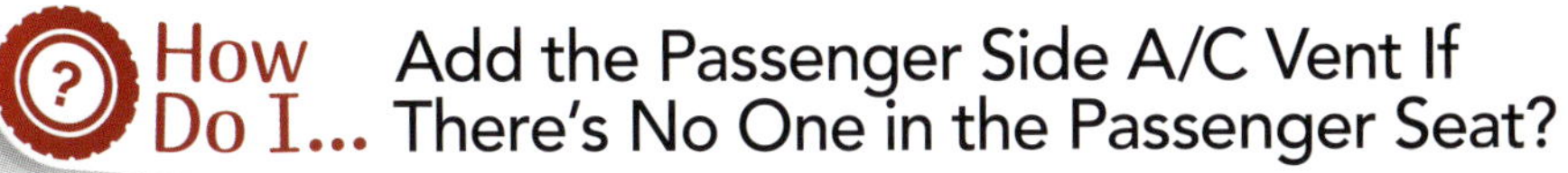

How Do I... Add the Passenger Side A/C Vent If There's No One in the Passenger Seat?

If there's no one sitting in the front passenger's seat of your car, the passenger side air vent is automatically turned off. However, if you want that passenger side vent to blow air anyway, in the bottom left of your center touchscreen, tap on the temperature reading to bring up the Climate Control screen. Now, tap on the passenger side vent where their airflow would come from and it turns on the airflow for that side of the car. If you change your mind and decide you don't want that extra airflow any longer, it's not as easy as just tapping on it again to turn it off (though that would make perfect sense if that's how it worked. Sadly, it doesn't—it's a little two-step dance to get it turned back off). To turn it off, bring up the Climate Control screen again, then tap on the floor airflow icon (it's to the right of the A/C button) to turn it on, then tap it again to turn if off, and now it's just back to blowing air on the driver's side. If that seems like a few unnecessary steps for just turning off the passenger side airflow, you're not wrong.

How Do I... Change the Fan Speed from My Steering Wheel?

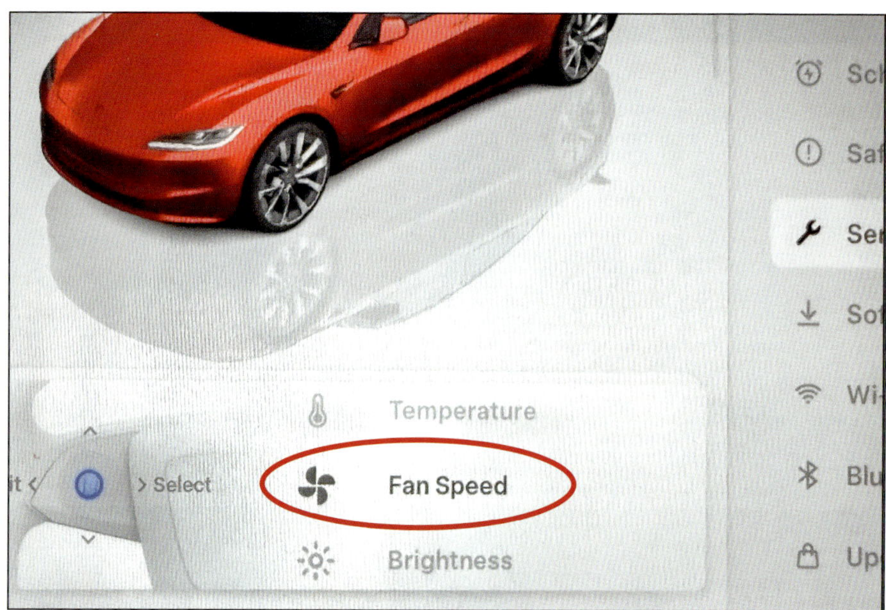

You can choose which features your left scroll button controls by tapping on the Controls icon (the car in the bottom left of your center touchscreen), then tapping on Display, and then going to Scroll Wheel Functions where you can choose from a list of features you can control (in our case, of course, we want it to control our fan speed). Another way to customize your left scroll button happens right on your steering wheel itself. Just press-and-hold the left scroll button, then look over at the bottom-left corner of your touchscreen and you'll see a pop-up menu (shown above) where you can scroll through the options of what this scroll button can control. In this case, use the left scroll button to scroll through your choices and when you get to Fan Speed, push it over to the right (it moves left and right) to make your selection. Now, you can control your fan speed using the left scroll button. By the way, the right scroll button is assigned to control Autopilot features (to see which ones, tap on the Controls icon, then tap on Autopilot, and you'll see which Autopilot features you can control with the right scroll button).

How Do I... Recirculate the Air?

In the bottom left of your center touchscreen, tap on the temperature reading to bring up the Climate Control screen, and then to the right of the fan icon, just tap the icon with an arrow going in a circle to recirculate the air.

TIP: QUICKLY ADJUST THE TEMPERATURE

Tap on the temperature reading at the bottom left of your center touchscreen and a slider appears, which you can drag right to raise the temp or left to lower it.

How Do I... Quickly Adjust the Climate Controls?

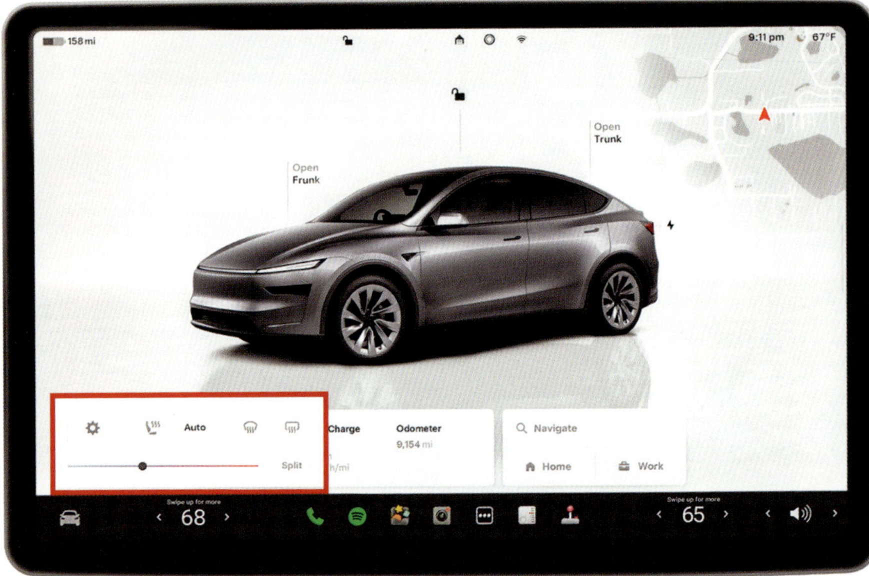

You don't have to bring up the whole Climate Control screen just to make some basic adjustments, like changing the fan speed, or using a slider to change the temperature, or to defrost the windshield, etc. To get to these controls quickly, down in the bar at the bottom of your center touchscreen, don't tap the temperature reading—instead, just tap once on the arrow on either side of it to bring up the Climate Control pop-up (seen here).

How Do I... Use the Rear Touchscreen Climate Control?

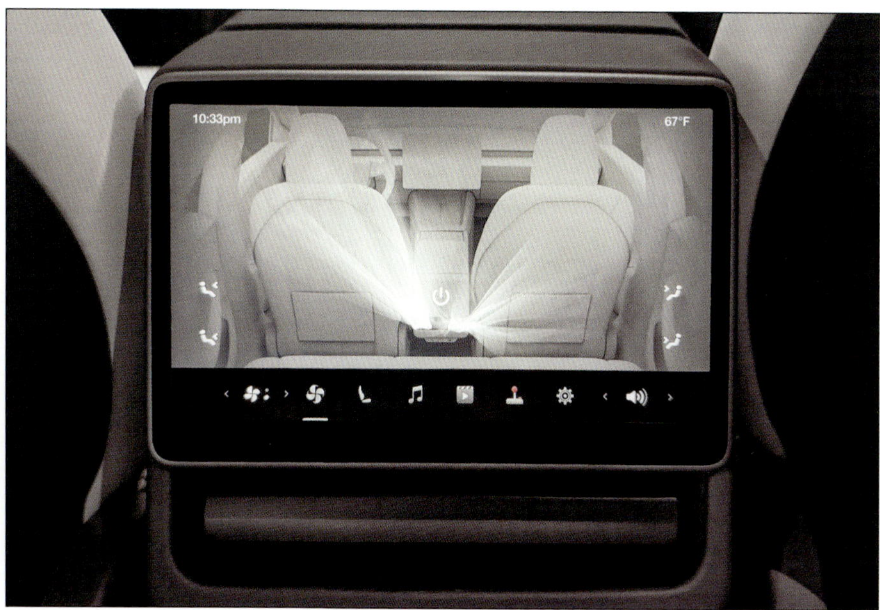

If you have a newer Tesla (and depending on your model), it may have an 8" rear touchscreen for the back seat passengers that is mounted on the back of the front armrest (the part that extends to the back seat). At the bottom of it, tap on the fan icon and there's a slider you can tap-and-drag to change the fan speed, there's an on/off icon, and you can control the direction and spread of the airflow from here, as well. There are also icons to choose whether you want the air to come from the vents below the touchscreen or the floor vents for the back seat, along with icons to turn on/off the rear heated seats. While the back seat passengers can control their own fan speed, and the direction and airflow pattern, they cannot change the temperature—that is only controlled from the main center touchscreen up front. *Note:* See page 52 for how to control the rear touchscreen from the front touchscreen.

How Do I... Turn On or Adjust the Front Heated Seats?

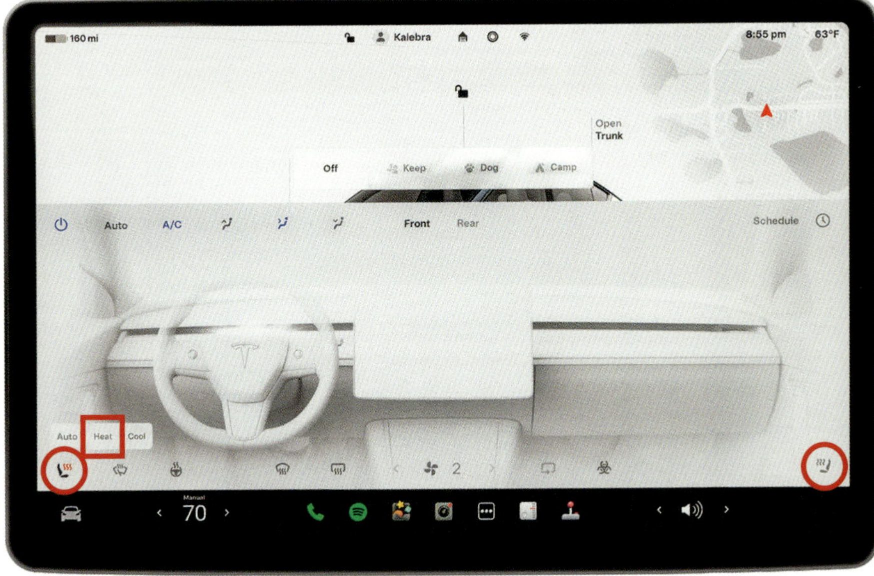

If you'll be using this feature a lot (i.e., you live in Minneapolis), you can add it to the My Apps area along the bottom of your center touchscreen, so it's always just one tap away. To do that, tap on the App Launcher (the three dots) at the bottom center of your touchscreen to bring up the app tray of icons you can add to your My Apps area (tap-and-hold on any one of them to see the full list). Tap-and-hold on the Heated Seats icon, and then drag-and-drop it down onto the bar at the bottom of your touchscreen to add it and now it's just one tap away. Another way to turn on the seat heaters is to tap either the temperature reading or the fan icon at the bottom of your touchscreen to bring up the Climate Control screen. Here, you'll see a seat icon on the bottom left for the driver's seat and another on the bottom right for the passenger's seat (they're both circled here). Tap either icon and a row of buttons pops up. Tap on Heat, and then tap on the seat icon again to turn the seat heating on at its maximum setting. If you want less heat, tap on it again, and for even less heat, tap on it a third time (you'll see the little red "heat" icon go from three red bars, to just two, and then down to just one for the lowest setting).

How Do I... Turn On or Preheat the Rear Heated Seats?

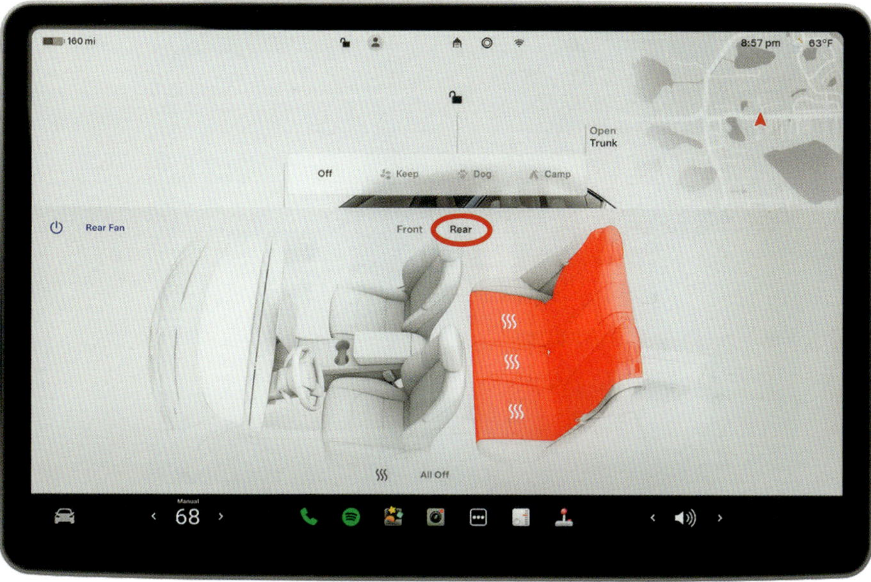

If you have heated rear seats, tap on the temperature reading in the bottom left of your center touchscreen to bring up the Climate Control screen and you'll see the Rear button grayed out. Tap that to bring up an illustration of the interior seats, then just tap on the rear seat you want to heat and that seat will turn red to indicate the seat heating is on. By default, it's at its maximum heat setting, so tap the seat again to lower it (the heat setting icons here are just like those for the front seats on the previous page). To turn all of the rear heated seats off, you don't have to tap on each one—just tap All Off at the bottom. Lastly, you can set these seats to automatically preheat before you get in the car by tapping Schedule in the Front Climate Control screen and choosing when you'd like them to heat up.

TIP: HEAT UP YOUR STEERING WHEEL & WINDSHIELD WIPERS

In the Climate Control screen, to the right of the driver's Heated Seat icon, are two other heating icons (if you have these features): tap the first one to turn on wiper defrosting and tap on the second one to heat up your steering wheel.

How Do I... Cool Down My Seats?

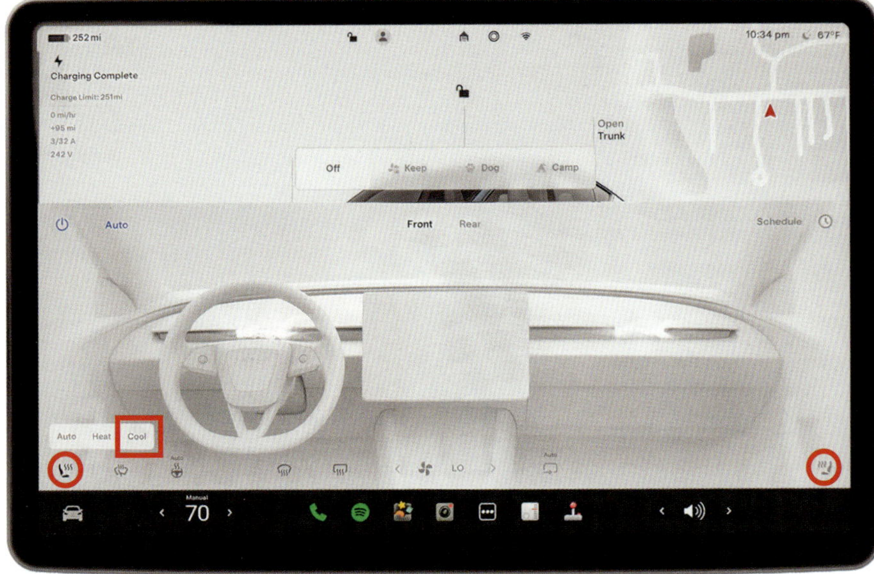

The newer Model 3 and Model Y come with ventilated seats that have a cooling option (like Models S and X have had for a while) to help you cool down your interior (well, at least your posterior). To turn this feature on, tap on the temperature reading in the bottom left of your center touchscreen to bring up the Climate Control screen. Now, tap on the seat icon in the bottom-left corner (for the driver, or the bottom-right corner for the passenger) and a row of buttons pops up. Tap on Cool, and then tap on the seat icon again to turn this on. If you have the seat heating/cooling icon down in your My Apps area at the bottom of your touchscreen (see page 86 for how to add icons down there), if you tap on the icon, a mini-card pops up in the bottom left of your touchscreen (below the rendering of your car) where you can just tap on Heat or Cool. *Note:* If you tap on Auto in the row of buttons, your front seats will automatically either warm up or cool down depending on the temperature in your car.

How Do I... Turn On My Window Defroster or Defogger?

In the bottom left of your center touchscreen, tap on the temperature reading to bring up the Climate Control screen. At the bottom, you'll find two controls for defrosting and defogging your windows (circled here): The one on the left is for your front windshield. Tap on it once to turn on the defogger (the icon will turn blue) or tap on it twice to turn on the defroster (the icon will turn red). Tap it again to turn it off. The one on the right is for defrosting your rear window. Just tap on it once to turn it on and it'll automatically shut off after 15 minutes (your Charge Port and side mirrors are also heated when you have this turned on). *Note:* You can also add either (or both) of these icons to your My Apps area at the bottom of your touchscreen (see page 86 for how to do that).

Chapter 4

How to Charge Your Tesla

There's More to It Than Just Plugging It In

Has this ever happened to you? You're talking with someone you've just met, maybe it's someone working on your house or a server at a restaurant, and when they notice you have a Tesla, they say something like, "You have a Tesla? I sure hope there's a charging station near you," but they say it with a tone like you're some kind of sucker. I can't tell you how many times I've heard that, or something close to it—and this is precisely why I carry pepper spray. Just a couple quick squirts and you can't believe how quickly they retract that statement. Okay, of course I'm kidding. I wouldn't use pepper spray on them—I'd use a taser instead. Tasers are much more effective, and the one I recommend for Tesla owners is the TASER Pulse 2, which fires projectiles for a longer range (nearly 15 feet), and has been "upgraded with improved durability and a new sleeker, textured design" with built-in laser-assisted targeting for enhanced accuracy. It can immobilize someone with an attitude working on your house (or a server at a restaurant) for up to 30 seconds, with what they call "neuromuscular incapacitation." Now, just a quick reminder: this is a chapter opener, so you can pretty much count on none of this being true. I do not have a TASER Pulse 2. I have the older, original one. See, I got you again. I've never had a taser of any kind, nor would I own one. I am a very non-violent person, except of course, when somebody asks me if I live close to a charging station and they have some "tone" in their voice. Okay, so here's what I do instead: I sucker punch them right in the face. No. I would never do that. But, you know, never is a strong word. Probably not (well, certainly not so far, but it's early in the day). Anyway, what I calmly tell them is that I have a charger at my house, plugged into a standard ol' dryer outlet in my garage, so I rarely ever need to visit a charging station, and that really surprises them. Then I ask them, "If you had a gas station at your house, how often would you fill up somewhere else?" They answer, "On a road trip." And I say, "Me too"—and then I add a colorful adjective that my publisher would not allow me to include in the book. Anyway, it's pretty convenient having that dryer outlet in your garage as long as you don't mind your clothes being kind of damp all the time.

How Do I... Know the Difference Between Level 1, Level 2, and Level 3 Chargers?

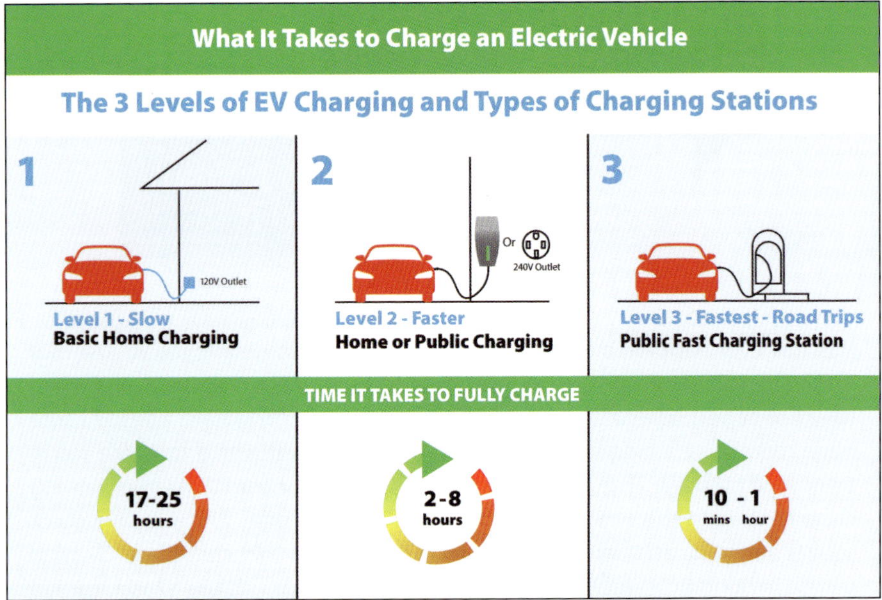

One of the first things you'll need to know is the difference between chargers and their speeds so that you can adjust your expectations. Electricity is electricity, but the speed at which your Tesla charges will vary by which kind of charger you use and its power source. Let's start with the standard outlet in every home or garage in the US. It's a standard 120v outlet (NEMA 5–15). Any charger you plug into this outlet is considered to be a Level 1 charger. It will be the slowest and you should probably only use Level 1 charging as a last resort. A Level 1 charger will only charge your Tesla at a rate of 4–6 miles of range per hour. So after being plugged in for 10 hours you can expect to add 40–60 miles of range. That's not a lot, but for someone who only drives 20–40 miles a day, it would work. The average Tesla owner goes with Level 2 charging. This is via a 240v outlet (NEMA 14–50), which is similar to the outlet that your electric dryer uses. Level 2 charging is much faster at a rate of up to 30–44 miles of range per hour. This means that most Teslas could fill up from empty in 10 hours or less. Lastly, there are Level 3 chargers. These are public chargers, such as Tesla's Supercharger network. They can recharge up to 200 miles in as little as 15 minutes. Level 2 chargers can be installed in your home and in public locations, such as parking lots, shopping centers, hotels, etc. In other words, places where you're likely to be spending some time. Level 3 chargers are not for your home. They are always installed in public locations and usually located near highways for people doing road trips. There could be a cost to charge at a Level 2 public charger, but some could be free. Supercharging does cost money, but is usually much cheaper than gas.

How Do I... Decide Which Charger to Buy?

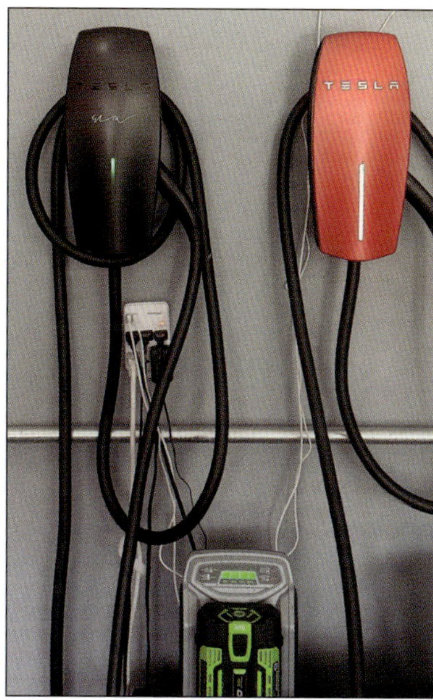

When you order your Tesla, you'll have the option to either order the Mobile Connector (about $250) or the Wall Connector (about $450–$580, depending on which model you want). What's the difference? As the name implies, the Mobile Connector is portable and designed to plug into an outlet either at home or on the go, such as a vacation property. Most Tesla owners go with the Mobile Connector, even if they just plan to leave it plugged in at home. With the Mobile Connector, you can use either a 120v or 240v outlet with the proper adapter (it comes with both). At home, you would have an electrician install a 240v outlet (NEMA 14–50) on a 60-amp breaker near where you park your car. One end of the Mobile Connector plugs into your wall outlet and other end plugs into your Tesla. You can leave it plugged into your power outlet even when you're not charging. The other option, the permanently installed Wall Connector, is nice because not only does it look better aesthetically, but it also charges your Tesla faster. The Mobile Connector does up to 30 miles of range per hour and the Wall Connector does up to 44 miles. If you're charging overnight, it probably won't matter which one finishes faster, but if you frequently come home and charge and go back out, then having the faster option could make a difference. As you might have guessed, with the Wall Connector you'll need a licensed electrician to install it on a 60-amp breaker. Also note that the Wall Connector is weatherproof and can be installed outside if you don't have a garage or a covered parking space.

How Do I... Charge at Home on a 120v Outlet?

If you opted to buy the Mobile Connector and haven't yet had an electrician install a 240v outlet (NEMA 14–50), you can charge your Tesla on a standard 120v household outlet (NEMA 5–15). Most garages have at least one on each wall and you can plug it into the one that is nearest the Charge Port on your Tesla. I highly recommend that you *do not use an extension cord.* Extension cords can cause all kinds of issues while charging your Tesla. First, they can simply not work because they aren't heavy duty enough. Second, even if they work, they will likely get very hot and can potentially cause a fire hazard. You are better off plugging your Mobile Connector directly into the outlet. Do not use surge protectors or power strips either. Since there is quite a bit of power being pulled, it's best if the outlet is not shared by anything else. In other words, unplug anything else that might be on the same circuit and only plug in your Tesla.

How Do I... Get a 240v Outlet (NEMA 14–50) Installed?

If you live in a property where you can have a dedicated outlet installed, this will be your best option. The cost will vary depending on how far your electrical panel is from where you park your Tesla. Luckily, my panel is right in my garage and my outlet was only $200 to install. However, if it's a long run, like one side of the house to the other, it could easily cost $2,000–$3,000. Those higher numbers are usually the worst-case scenario. My friends have paid on average $200–$600. Make sure you tell your electrician that you need a 240v NEMA 14–50 on a 60-amp breaker. While you could go with a smaller breaker, it will impact the speed at which your Tesla charges. Going with a 60-amp breaker will allow the Mobile Connector to charge at its fastest speed and the cost between breakers is negligible.

How Do I... Use Other 240v Outlets Besides a NEMA 14–50?

If you are having a 240v outlet installed to charge your Tesla, you should definitely get a NEMA 14–50, which is the current standard. However, there are other 240v outlet styles, especially in older homes or public locations, such as campgrounds. Luckily, Tesla sells a set of 240v plug adapters for your Tesla Mobile Connector. Although I didn't have a need for these, I still bought the set and keep it in my lower trunk just in case I'm in a situation where I need one and it's more convenient than driving to a Supercharger—I'm thinking grandma's house. The NEMA Adapter Bundle sells for $245 via the Tesla online store (https://kel.by/nemabundle) and includes one of each of the following adapters:

- Gen 2 NEMA 5–20 Adapter
- Gen 2 NEMA 6–15 Adapter
- Gen 2 NEMA 6–20 Adapter
- Gen 2 NEMA 10–30 Adapter
- Gen 2 NEMA 14–30 Adapter
- Gen 2 NEMA 14–50 Adapter
- Gen 2 NEMA 6–50 Adapter
- Storage Bag

How Do I... Get the Tesla Wall Connector Installed?

Although I started with a Mobile Connector and a 240v outlet (NEMA 14–50) with my first Tesla, I quickly learned that I wanted a better-looking charging setup. I wanted a permanent install. So, I ordered a Tesla Wall Connector from Tesla and had my electrician come back out and install it. Not only did it look better, but it charged my Tesla faster, and it can also be installed outside as it is weatherproof. The latest model has WiFi built in, which allows you to control it via the Tesla app, as well as do software updates on it. You can also restrict access to it via an approved list of VINs. (This is great if you do install it outside and you don't want random strangers trying to charge while you're not home.) You'll also want to have it installed on a 60-amp breaker, so that it can charge at its maximum speed. They even sell different colored covers for it so that you can match the color of your Tesla. The charging cable conveniently wraps around it and the charging plug goes in the right side of it when not in use. They sell two models: the base model has a North America Charging Standard (NACS) plug on it to match the one built into every US-made Tesla, and the more expensive Universal model actually has both a NACS plug and a J1772 plug/adapter built in. This way you can not only use it to charge a Tesla, but you can use it to charge any other brand of EV as well, making it a great option for multi-brand EV households.

How Do I... Know How Often I Need to Charge My Tesla?

How often you charge your Tesla usually depends on how often and how far you drive day to day and whether or not you have the ability to charge at home or at work. I work from home and don't have a daily commute, therefore, I go some days without even driving my Tesla. However, when I come home, I plug it in and leave it plugged in. This way, I always wake up to an 80% charge. I never have to think about stopping at a charger when I'm out and about each day. Sometimes I do forget to plug it in, so this means that the next time I go out I may be down to a 60% charge or a 50% charge, which is still plenty to get me everywhere I need to go that day. While you certainly don't have to charge every day, I can't think of any reason not to. Plugging your Tesla in when you come home for the night should just be a habit. This way, you never have to think about charging. If you don't have a home/work charging setup, then you'll need to plan to charge at public chargers when it's convenient for you. Usually people that don't have home/work charging setups charge up once or twice a week, depending on their daily commute.

How Do I... Open and Close My Charge Port?

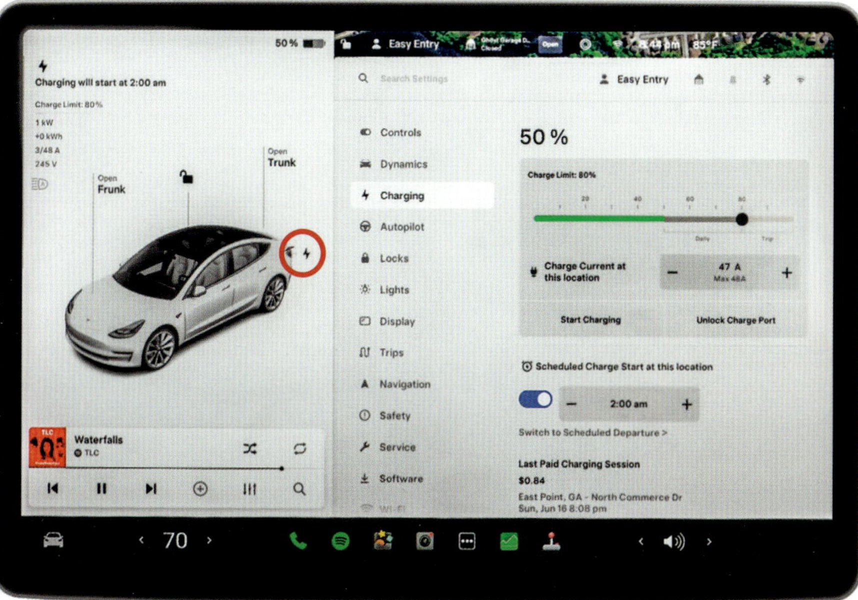

Your Tesla's Charge Port is located on the driver's side near the taillight. You can open it a few different ways: Usually, you'll be standing right next to it and want to open it because you're ready to plug in your charger. If you're using a Tesla-branded charger or are at a Supercharger, there is a button right on the top of the charging cable. If you press it, that will open the Charge Port door. If for whatever reason that button doesn't open it, you can tap the door with your hand and that should open it. You can also open the Charge Port door from the Tesla app, and while you're inside your Tesla, you can open it from your center touchscreen by tapping on the little lightning bolt icon near the rear driver's side (seen circled above). The Charge Port door will close automatically after a few seconds if there is no charger plugged in. Never try to force it closed with your hand—it doesn't work that way.

How Do I... Plug In and Unplug My Tesla?

Once the Charge Port door is open, all you have to do is push the charger all the way in, and the indicator light on the side of the Charge Port will change color to let you know that your Tesla is plugged in. If your Tesla is set to start charging immediately, the light should start flashing green to let you know that it is, in fact, charging. The Tesla charger and the J1772 and CCS adapters all lock into place once they are plugged in, which prevents someone from either unplugging your Tesla or stealing your adapter. To remove the charger, press-and-hold the button on the top of the charging cable to unlock the Charge Port, and then you can simply and safely pull the charger out. You can also unlock it by tapping the Unlock Charge Port button in either the Tesla app or on the Charging screen (tap on the Controls icon—the car in the bottom left of your center touchscreen—then tap on Charging). If it hasn't finished charging yet, it's a good idea to tap the Stop Charging button first (in the app or on the Charging screen). If you're using a generic charger with an adapter, it's best to pull the charger and adapter out at the same time, and then remove the adapter from the generic charger.

How Do I... Know If I Need to Unplug My Tesla After It Finishes Charging?

This question comes up all the time with new Tesla owners. Thankfully, your Tesla's charging hardware is really smart. Once your Tesla reaches the charge limit that you set (see page 107), it will automatically stop charging. There is no need to unplug it until you're ready to go. As a matter of fact, Tesla recommends that you leave it plugged in at all times, if possible. As I mentioned, I leave my Teslas plugged in all the time when I'm not driving them. This way, I also don't forget to plug them in because it's a habit when I come in for the day.

How Do I... Charge at a Tesla Supercharger?

At some point, you're likely going to charge at a Supercharger, whether you're going on a long road trip or you forgot to charge at home or work and need a little more juice to make it to your destination. As of the writing of this book, there are over 50,000 Superchargers worldwide, so chances are there is one near you or where you're going. If your state of charge is too low to make it to your destination, your navigation system will automatically route you to a Supercharger that is along your route. Once you arrive, just back into an empty stall (preferably leaving one between you and the next Tesla—more on that later). The Charge Port is located to the rear of the driver's side on every Tesla. Once you plug in, your Tesla should start charging within a few seconds. There is nothing else you need to do to initiate the charge, and your credit card on file will be billed for the cost of charging. You'll see your state of charge right on your center touchscreen as it charges up. Unlike a gas car, there is rarely a need to "fill up" (charge to 100%)—you only need to charge enough to make it to your destination or next charging stop. The lower your state of charge when you pull in, the faster your Tesla will charge, and as your battery gets near full, it will charge slower. Usually it's a literal waste of time to charge past 80%. When you're done charging (it's important to stop the charge via your Tesla's touchscreen or the Tesla app), you'll want to unplug and move your Tesla so that the next person can use the Supercharger. Also, if you leave your Tesla plugged into a Supercharger and it's finished charging, Tesla will likely start charging you idle fees, so charge, and then unplug and move your Tesla. (*Note:* See Chapter 9 for more on using Superchargers.)

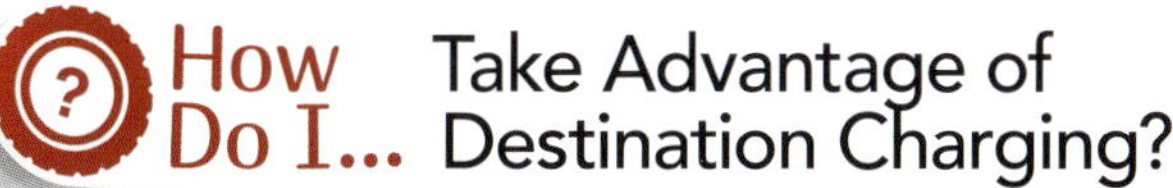

How Do I... Take Advantage of Destination Charging?

Your Tesla navigation system assumes that you will be able to charge, if necessary, once you reach your destination. It's up to you to charge longer along the way on a road trip if you want to arrive with a higher state of charge to be able to get around once you arrive. However, in most cases, I'm either arriving with plenty of charge to make the trip back home or I'm planning to charge at my destination. When I'm on a road trip I usually pick hotels that offer EV charging. This is like being at home. When I arrive at the hotel they either have chargers in the parking lot or they offer charging if you valet park and they plug you in. This is awesome because I don't have to think about charging and can just enjoy my stay. The best part is that at every hotel I've stayed at to date, the charging was *free*. Free charging may not always be the case, especially in crowded cities like New York or San Francisco, but in most smaller hotels across the country, EV charging is a free perk or no more than the cost of parking. Remember to have your J1772 adapter handy, as most destinations have generic chargers so that they can charge any EV.

How Do I… Use an Adapter (J1772 or CCS) to Charge at a Non-Tesla Charger?

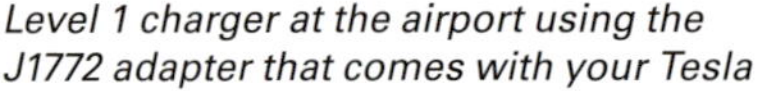

Level 1 charger at the airport using the J1772 adapter that comes with your Tesla

Level 3 Electrify America charger using the optional CCS adapter

While all the top EV manufacturers have now adopted the NACS (Tesla) charging port on their future EVs, there are still quite a few chargers out there that use the older standards of J1772 (Level 1 and Level 2) and CCS (Level 3). Level 2 chargers from companies like ChargePoint, Shell Recharge (formerly Volta), and Blink have tons of chargers around town that have J1772 plugs on them. Luckily, Tesla includes a J1772-to-Tesla adapter with your Tesla. You can put this adapter on the end of the plug from a generic Level 1 or 2 charger and plug it into your Tesla. Depending on the charger, you may have to download their app and pay via the app or a touchscreen on the charger itself. Most hotels that provide EV charging will have these generic Level 1 and 2 chargers and you'll need your adapter to use them. If you want to be able to use third-party Level 3 chargers, like those at Electrify America, you'll need to buy a CCS Combo 1 adapter (Tesla sells one for $250 for newer Teslas). Like the J1772 adapter, you'll attach it to the end of the generic Level 3 CCS plug and then plug it into your Tesla. Like the Level 1 and 2 chargers, you'll have to either download an app for that charging network to start the charge or use the built-in touchscreen on the charger to start and pay for the charge.

How Do I... Charge for Free?

In all my years of driving, I've never once gotten free gas. However, there are free chargers in lots of places. There is even an ad-based charging network called Shell Recharge (formerly Volta) that puts chargers in shopping center parking lots. The idea is that the advertisers pay for the charging by displaying ads on the big screens of the chargers. You just pull up and plug in via your J1772 adapter. These are Level 2 chargers and typically slow, but free is free. If you're going to be shopping for an hour or dining at a restaurant, you could charge enough to replace the energy you used getting to and from the shopping center, effectively making the trip free (except for the shopping you do—that's on you!). More and more parking lots are including free EV charging for the cost of parking, which you were going to pay anyway. For example, my local airport offers free charging in their parking lots. These are Level 1 chargers, which are extremely slow. However, since you're probably going out of town for at least a day or two, your Tesla will be fully charged by the time you return.

How Do I... Precondition My Battery?

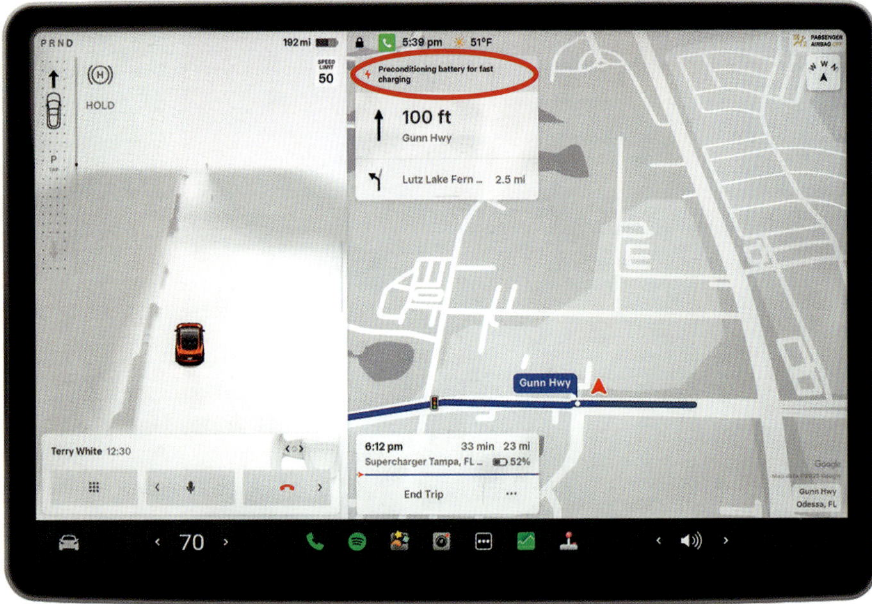

Your Tesla can charge even faster at a Supercharger if you precondition your battery first. This is actually something that happens automatically when you navigate to a Supercharger via your Tesla's navigation system. It sees that you're headed to a Supercharger and therefore starts to warm the battery for a faster charging session when you arrive. This is also extremely useful when driving during the winter in colder climates. Even if you didn't precondition your battery, your Tesla will still Supercharge just fine, but navigating to the Supercharger via your Tesla navigation system—even if you know how to get to it—will save you time with a faster charge.

How Do I... Set My Charging Limit?

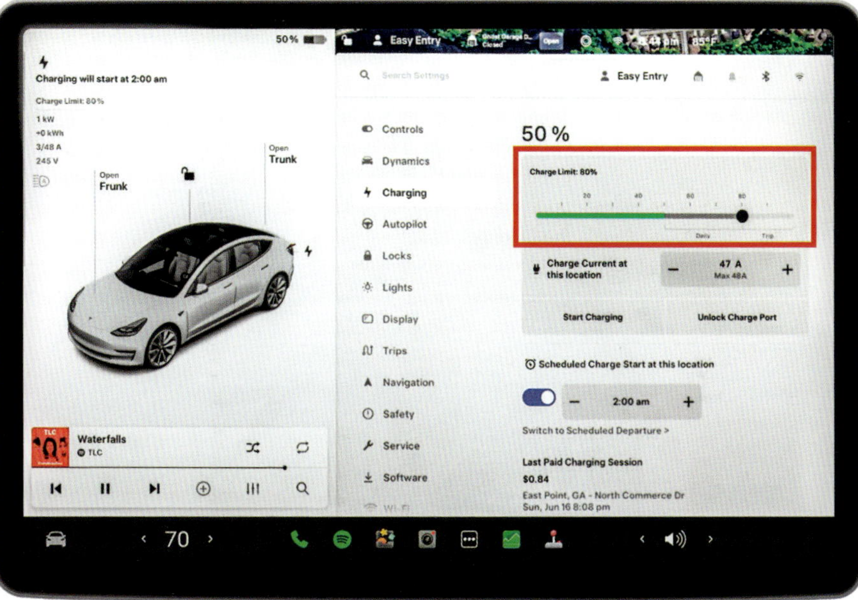

We've driven gas vehicles for years and as long as we had enough money, we could fill it up when we stopped for gas. Why? The only reason was so that we wouldn't have to stop for more gas until it was nearly empty again. This is ingrained into our psyche. However, when it comes to a Tesla/EV, the rules are a bit different, especially if you have the ability to charge at home. First of all, for most Teslas, Tesla recommends only charging to 80% day to day. But, if you have a Tesla with newer LFP batteries, then Tesla actually does recommend charging to 100% each time. When you're on a road trip, your Tesla navigation system will automatically route you to Superchargers along your route, if needed, and you'll notice that, in most cases, it will only have you charge up enough to make it to the next Supercharger. So, instead of one long stop, you make more shorter stops. But, there's nothing stopping you from charging longer at a Supercharger if you have the time (like I mentioned on page 102). To set your charging limit, tap on the Controls icon (the car in the bottom left of your center touchscreen), then tap on Charging, and then set it using the Charge Limit slider.

How Do I... Avoid Going Down to a 0% Charge?

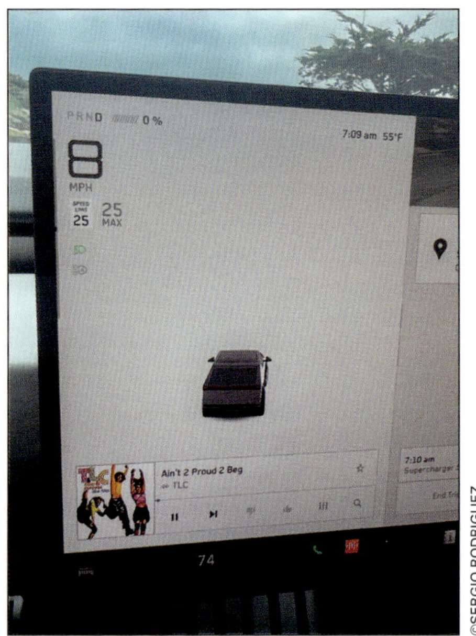

©SERGIO RODRIGUEZ

My friends and I have come up with the term "Club Static" for when you let your EV get all the way down to 0% (or 0 miles), but make it to a charger before it shuts off. I've never had the guts to let my Tesla get this low. It's also not good for the battery to let it get this low. Your Tesla will go out of its way to give you multiple warnings so that this never happens to you and, quite frankly, you'd have to completely ignore all the warnings to run out of charge and need a tow. However, most of the warnings happen only if your Tesla knows where you're headed. This is why it's always best to use your Tesla's navigation system even if you know the way. Your Tesla will always calculate if you have enough charge to make it or navigate you to a Supercharger along the way. Your Tesla will also give you suggestions to extend your range such as driving at a slower speed.

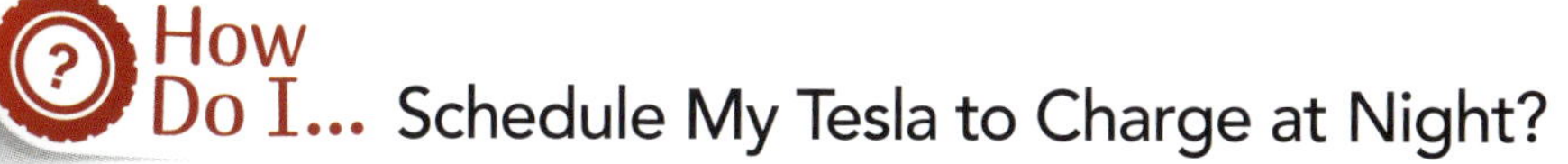

How Do I... Schedule My Tesla to Charge at Night?

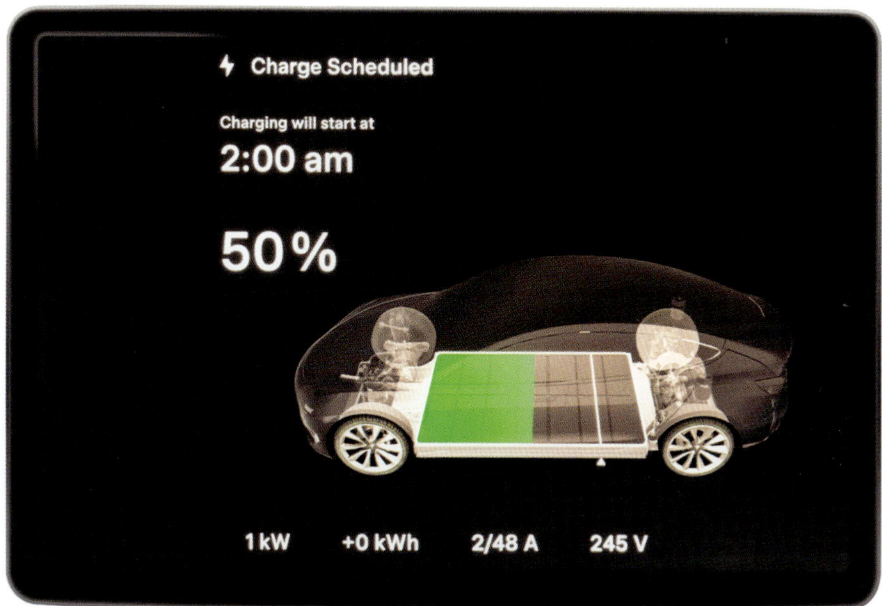

There are a couple of benefits to scheduling your Tesla to charge at night. The first one is that it could actually save you money if your utility offers an EV rate plan (as I describe on page 111). The second benefit is that it actually helps out the grid. Our electric grid has the highest demand during summer days. As long as you wake up to a full (or 80%) charge in the morning, it really doesn't matter that it charged in the middle of the night vs. right when you plugged it in. You can change the time your Tesla charges by tapping on the Controls icon (the car in the bottom left of your center touchscreen), and then tapping on Charging and setting your start time. You can also do this right within the Tesla app. For example, with my utility's EV rate plan, my super-off-peak time is from 11:00 p.m. to 7:00 a.m., and I have my Tesla set to start charging at 12:30 a.m. That gives it plenty of time to charge up to the limit that I set, which is 80%, by 7:00 a.m. each day, even if it was on 0% (I would never allow it to get that low). Another strategy for Tesla owners who have solar panels or solar roofs is to charge during peak solar production times during the day. You can only schedule one time, but it can be whatever time is most beneficial for your situation. Also note that you can override the schedule whenever you need to by tapping Start Charging on the Charging screen, or in the Tesla app, which immediately starts the charging process.

How Do I... Minimize Battery Drain While My Tesla Is Parked?

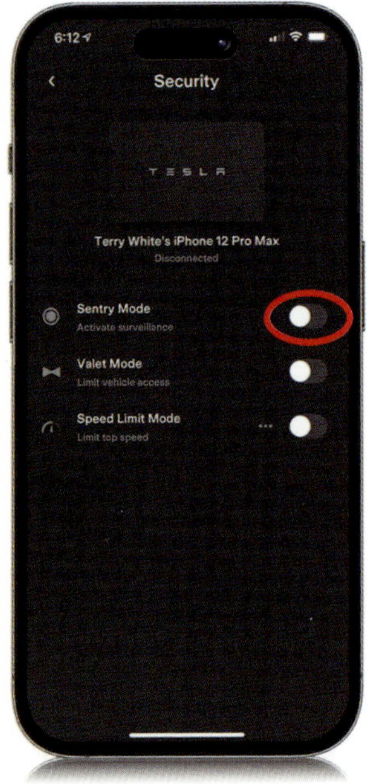

Any time you park your Tesla and it's not plugged in, your battery will drain a little because the systems that run your Tesla use a little power. This drain will be even more if you use Sentry Mode, Dog Mode, or Camp Mode. Each time you launch the Tesla app to check on your Tesla, it wakes it up and uses a little battery. Normally, this isn't a big deal if you have plenty of charge and you're not leaving it parked for a long time. However, if you are going on a trip for a couple of weeks and leaving your Tesla in a parking lot at the airport (unplugged), then this is more of a concern. The best practice is to arrive at the location that you're going to leave your Tesla parked at for an extended time with as high a state of charge as you can. Turn off Sentry Mode unless you feel that you want the security features that it offers (see page 179 for more on Sentry Mode). If you are going to leave Sentry Mode enabled, then you should try to park in an area with as little traffic/foot traffic as possible, as your Tesla will wake up anytime there is activity near it. While it's tempting to check in on your Tesla via the app while you're away, remember that each time you do, it wakes it up and drains the battery a little more. Don't open the Tesla app unless you really need to—enjoy your trip! You should have plenty of charge when you return.

Get an EV TOU Rate with My Utility?

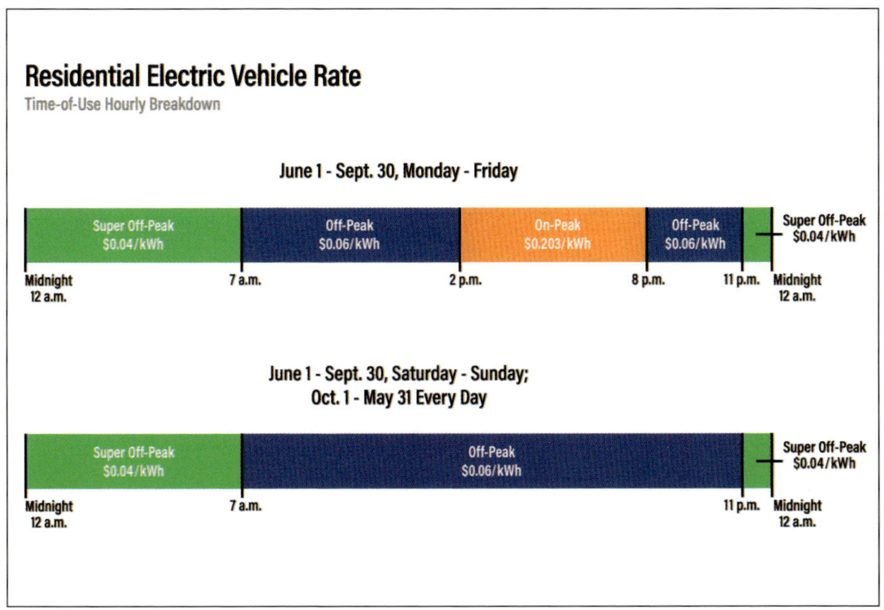

Electricity is usually in higher demand during the day as people use their air conditioners at home and in work locations. Utilities would love for you to charge your Tesla during non-peak times, like in the middle of the night when the demand on the grid is a lot less. Most utilities offer an EV plan or a Time of Use (TOU) plan that offers the lowest rates during "super-off-peak" times. The downside is that they charge a higher rate during peak times. As long as you can shift your heavy electrical needs to off-peak times, you can save quite a bit. Not to mention that these peak times are usually only Monday through Friday during the summer months. On the weekends, you can usually get off-peak and super-off-peak rates all weekend long. When I come home, I plug in my Tesla; however, charging isn't scheduled to start until after midnight at the cheapest rates (see page 109 for more on scheduling). Check with your utility to see what your EV rate plan options are.

How Do I... Find Non-Tesla Branded Chargers?

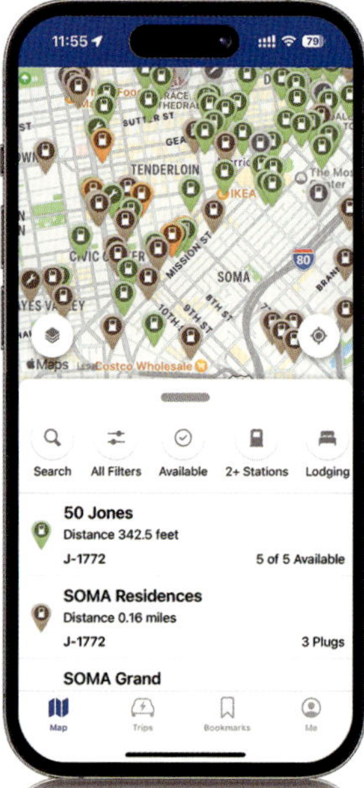

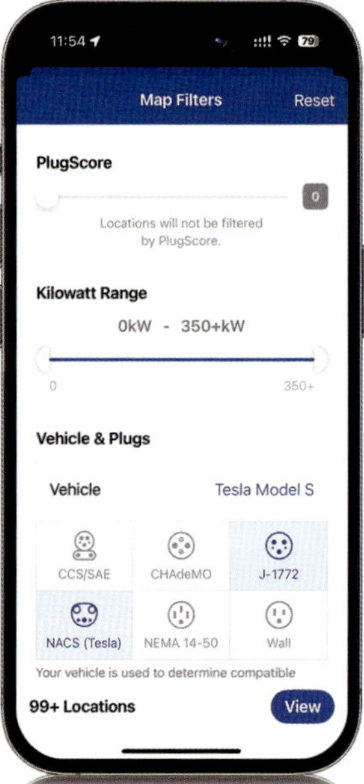

Both your Tesla app and your Tesla's navigation system do a great job of showing you nearby Tesla-branded Superchargers and Destination Chargers, but there may be chargers that you could use that are closer or more convenient than those listed in your Tesla app or navigation system. There's a great free mobile app for your smartphone called PlugShare. PlugShare shows all nearby chargers and you can even put in an address or area where you're headed to see the chargers near your destination that you can take advantage of. I often use PlugShare to see which hotels at my destination have on-site EV charging. It definitely impacts which hotel I choose. If a hotel has a charger on-site and I can wake up each day of my trip to a full charge, that's a win! You can even filter by the type of plug they have so that you can narrow the search down to those chargers that either have NACS (Tesla) plugs or the ones you have an adapter for, such as J1772 or CCS/SAE. Search your app store for the PlugShare app and download it.

How Do I... Display My Charge as a Percentage or Miles?

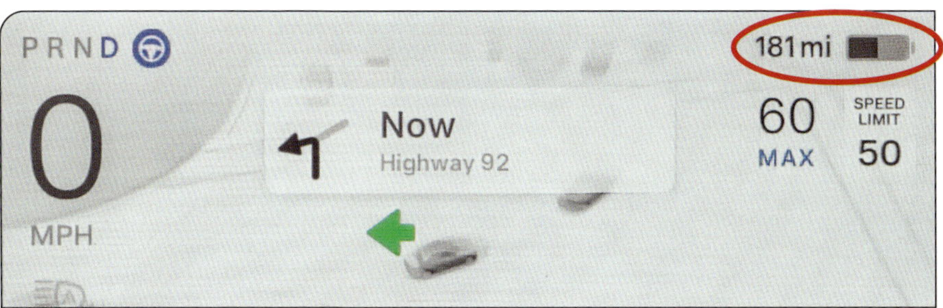

It's a preference! This can actually turn into a heated debate among Tesla owners. You can switch your Tesla's battery display to be either a percentage or the estimated number of miles left. I've heard the arguments for both. The people that usually pick percentage do so because they say the mileage estimate is very inaccurate. In other words, if it says you have 200 miles of range left, that is highly dependent on several factors, like how you drive, how fast you drive, weather, wind, temperature, etc. So that 200 mile number is likely less in reality. Therefore, why use miles? The people that favor the miles display will tell you that they would rather have a guess than have to do math with a percentage. I actually use both! I use percentage day to day because I don't need to see a miles estimate to go grocery shopping. However, on a road trip, I switch to miles every time because I want an idea of how far I could go before needing to charge. You can switch between miles and percentage right on your center touchscreen by tapping the miles or percentage next to the battery icon, and it will switch to the opposite one. If you have a Model S or Model X, you'll need to make the change on the Energy Display screen (tap on the Controls icon—the car in the bottom left of your center touchscreen—and then tap on Display).

Chapter 5

How to Use the Entertainment System

This Is Where It Gets Fun

I remember the first time I ever saw Tesla's entertainment system on that big bright touchscreen in the center of the car. I was blown away. Blown away! Up until that point, there were no big screens in cars, and you were lucky to have an FM radio or a cassette player as your entertainment—maybe SiriusXM, if you were really lucky. But now, we've got so many choices for music and video streaming that we just take it all for granted. Back in my day, if we wanted entertainment in our cars, we'd sit in the parking lot of a McDonald's with a small block of pine and we'd whittle away at it with a pocket knife. It was the technological equivalent of using a butter churn. It was painful. You got blisters and splinters, and you had to walk up hill both ways just to use it. After a while, you'd think about praying for one of those scary hovering death robots to zap you out of existence, but you didn't do that because you knew one day there would be a big ol' screen in your Tesla and you could watch YouTube as you sit in the McDonald's parking lot. Today, we use a gratuitous number of music and streaming sources in our Teslas just because we can. You brag about how many streaming services you subscribe to while sipping a glass of Beaulieu Vineyard Georges de Latour Private Reserve Cabernet Sauvignon, with one eye in the mirror as you watch yourself gavotte. You spend your evenings at fancy Tesla parties because once you got a taste of that "streaming music and video in your car" life, you left the rest of us behind to pick up the pieces of your shattered soul. Now you're just a vacant shell of a human being, flitting from social event to cocktail hour to sampling canapés at the country club, while casually dropping names of Tesla influencers you saw on the slopes last time you were skiing in Davos. "Oh, did I mention I saw Marques at the bar at Balthazar?" you say in passing, as you sample the rumaki while noting aloud it tasted a bit "chewy." Yes, once all these streaming and music choices came into your life, everything changed—not just for you, but for those around you. Poor souls still toiling in FM radio and dreaming of a more fulfilling life. You want to enlighten them. To pull them up to your level and give them a taste of "the streaming life." You could easily do it, but you know you never will. Fade to black. Roll credits.

How Do I... Quickly Listen to Music in My Tesla (LiveOne)?

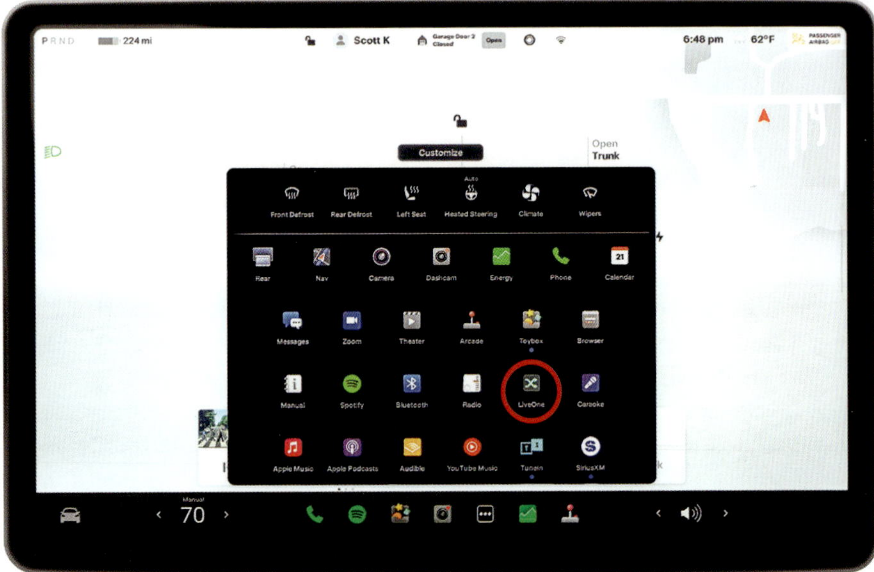

The quickest way to start listening to music in your Tesla right now is to press the Microphone button (or the right scroll button in older models) on the right side of your steering wheel, then release it and just say the song you want to hear. For example, say "Play the song 'California Girls' by Katy Perry," and in about two seconds, it starts playing. What you're using is a streaming service called "LiveOne" (formerly known as Slacker Radio), which is an Internet-based radio station that is surprisingly good, and they have almost any song you can imagine. Say, "Play 'Witchcraft' by Frank Sinatra," or "Play Vivaldi's 'Double Concerto for Strings'"—whatever you can think of—and it'll play it (it's hard to stump it). The nice thing about LiveOne is that, like your built-in navigation system, and YouTube, and browsing the Internet using your car's built-in web browser, you don't need a Premium Connectivity subscription (currently $10 a month) to access it (it comes with the Standard Connectivity package, which comes free with your Tesla. Well, you actually paid for it because you paid for the car, but you know what I mean). The long and short of it is, this is a fast, easy, free way to listen to the songs you want and it's better than some of the other paid options out there.

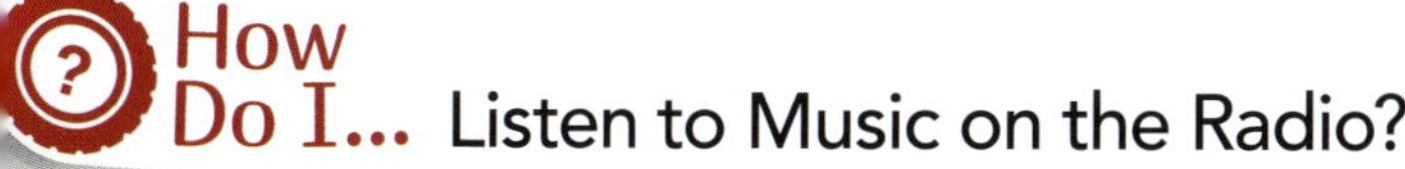

How Do I... Listen to Music on the Radio?

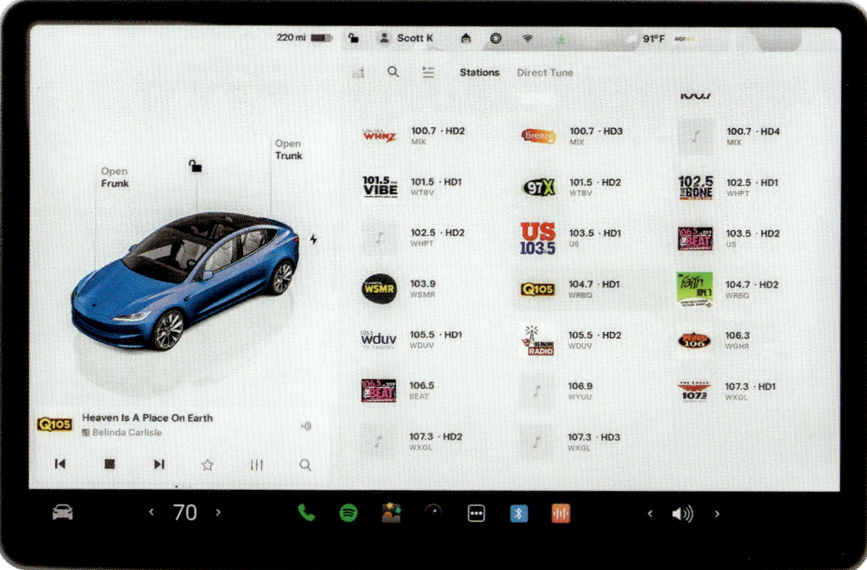

Tap on the App Launcher (the three dots) icon in the bottom bar of your center touchscreen to bring up the app tray and you'll see all sorts of streaming sources, like Spotify, Apple Music, SiriusXM, and so on (more on those on the next page). To listen to the radio, tap on (wait for it...wait for it...) Radio. Teslas only have FM radio, but if you really want to listen to AM, one thing you can do is go to TuneIn (it's in the app tray and is primarily a podcast app) and tap on one of the AM stations listed there.

How Do I... See My Music Streaming Options?

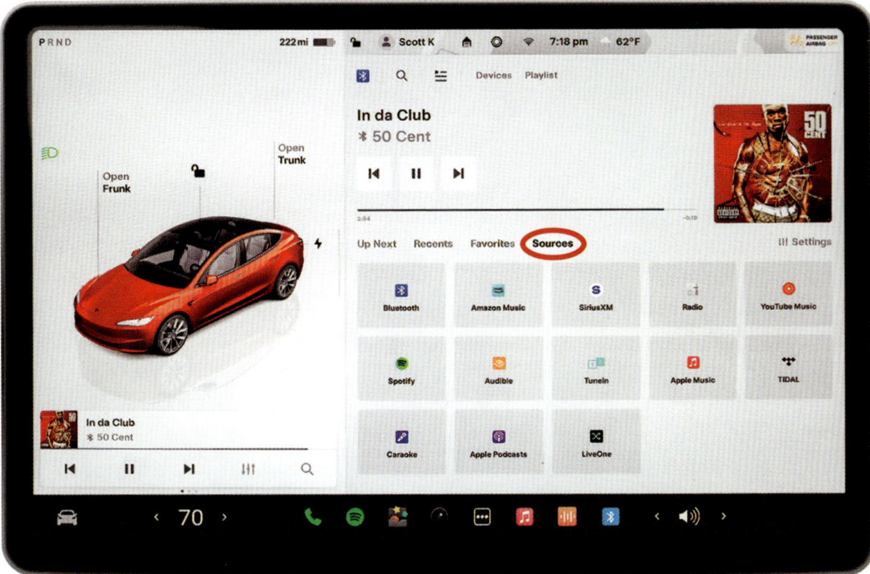

First, here's a quick shortcut: You know that little mini-Music Player card that appears at the bottom left of your center touchscreen when you're playing music (the one under the image of your car)? Tap on it and on the right side of your touchscreen the Music Player appears. Near the top, you'll see tabs for Up Next, Recents, and Favorites, but the one you want to tap on is Sources. This brings up tiles for all your music streaming sources (as seen above) and you can tap on the one you want right there. Another way to get here is to tap on the App Launcher (the three dots) icon in the center of the bottom bar of your touchscreen, and then in the app tray, you can tap on any one of the built-in streaming music services shown there, including Spotify, Apple Music, TIDAL, Apple Podcasts, Audible, TuneIn, and LiveOne (which used to be Slacker Radio; see page 116 for more on this one).

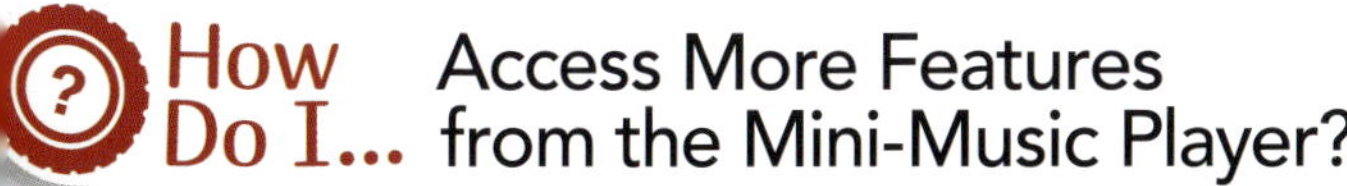

How Do I... Access More Features from the Mini-Music Player?

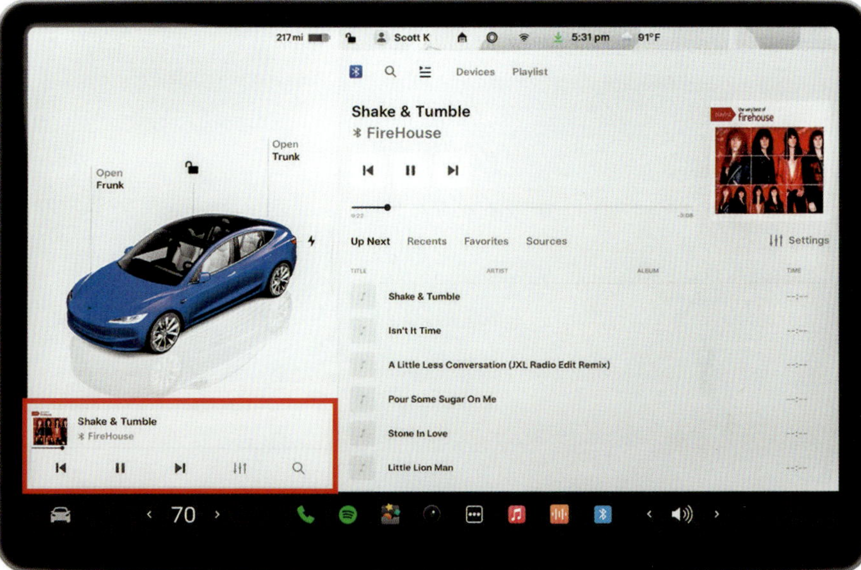

If you're playing music, a little mini-Music Player card pops up at the bottom left side of your center touchscreen where you'll see play/pause, fast forward, and reverse icons. Swipe it up and you'll also see shuffle and repeat icons, and a heart icon for marking a song as a favorite. To get to more features, just tap on that card and the Music Player appears on the right to reveal things like recently played albums, a repeat button, a button to take you to the EQ controls, and a search field (you tap on the magnifying glass icon).

How Do I... Listen to Songs I've Loaded onto a USB Flash Drive?

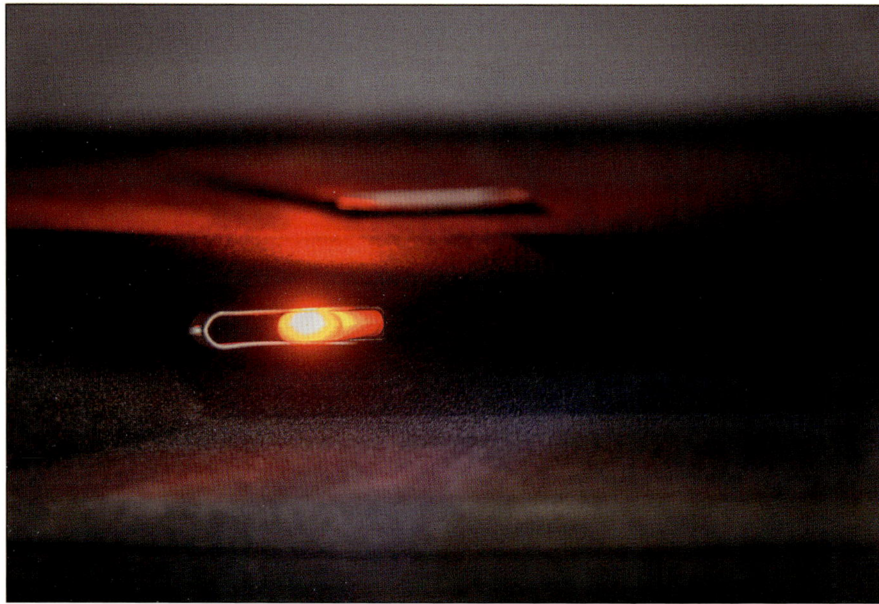

To play music you've loaded onto a USB flash drive, first insert your USB flash drive into a USB port inside your front center console (USB-C) or inside your glove box (USB-A). Once it's inserted, you should see the USB icon appear in the bottom bar of your center touchscreen. Tap on it and the Music Player will open. If you have your music in folders, tap on the one you want, and then just tap on a song to begin playing it. (*Note:* Some newer models have USB-C ports that only support charging, so you may need to try using the USB-A port in your glove box, if you're not seeing the USB icon. Also, check the book's companion webpage, mentioned in the book's intro, for some other troubleshooting ideas, if you have a problem accessing your USB flash drive.)

TIP: HOW TO FORMAT YOUR USB FLASH DRIVE

Insert the drive into a front USB port in your Tesla, then tap on the Controls icon, then tap Safety, and then tap Format USB Drive. This will automatically format it as exFAT (which is the one that Tesla recommends).

How Do I... Connect Someone's Phone So They Can Play Music in My Tesla?

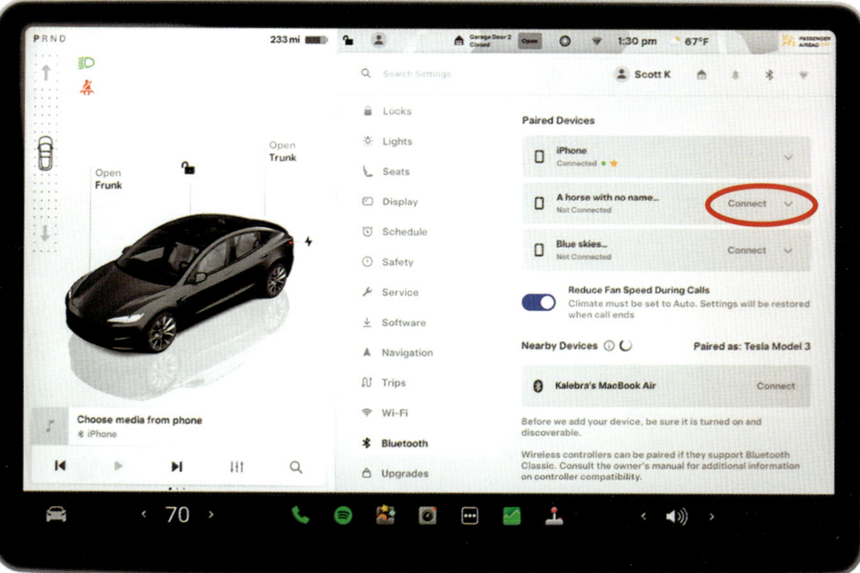

This is actually pretty simple—just have them connect to your car via Bluetooth. Tap on the Controls icon (the car in the bottom left of your center touchscreen), and then tap on Bluetooth. Make sure your passenger has their phone unlocked, has Bluetooth turned on, and has it in Discover mode, so your Tesla can find it. Once it does, you'll see the name of their phone appear beneath Paired Devices. Tap Connect to the right of its name and your Tesla will display a number asking if that's the same number they see on their phone (just to make sure you don't connect some freaky stranger's phone by accident). If the numbers match, tell them to tap on Pair on their screen, and once they're connected, they're all set. Now, when they play music on their phone, it will play in your Tesla.

How Do I... Adjust the Volume?

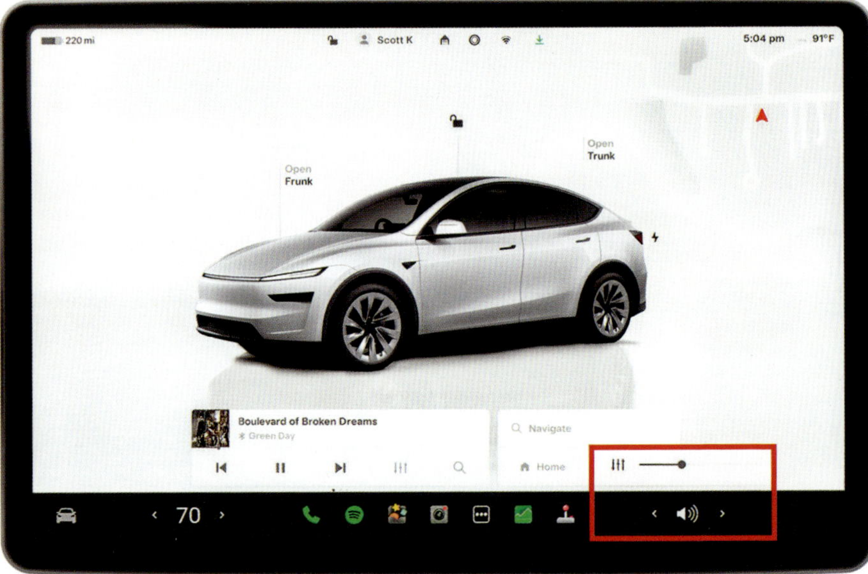

There are a few ways: My favorite way is to tap-and-hold on the speaker icon in the bottom bar of your center touchscreen and a pop-up volume slider appears. Adjust the volume by just sliding your finger to move the slider. Another option is to tap the left and right arrows beside the speaker icon—tapping the right arrow raises the volume; tapping the left arrow lowers it. You can also use a voice command to raise or lower the volume by pressing the Microphone button (or the right scroll button in older models) on the right side of your steering wheel, then releasing it and saying, "Volume up," "Volume down," or even "Mute audio." Lastly, you can use the left scroll button on your steering wheel to control the audio volume.

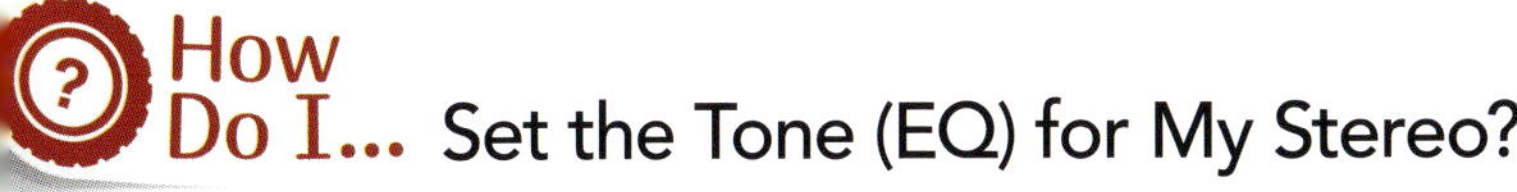

How Do I... Set the Tone (EQ) for My Stereo?

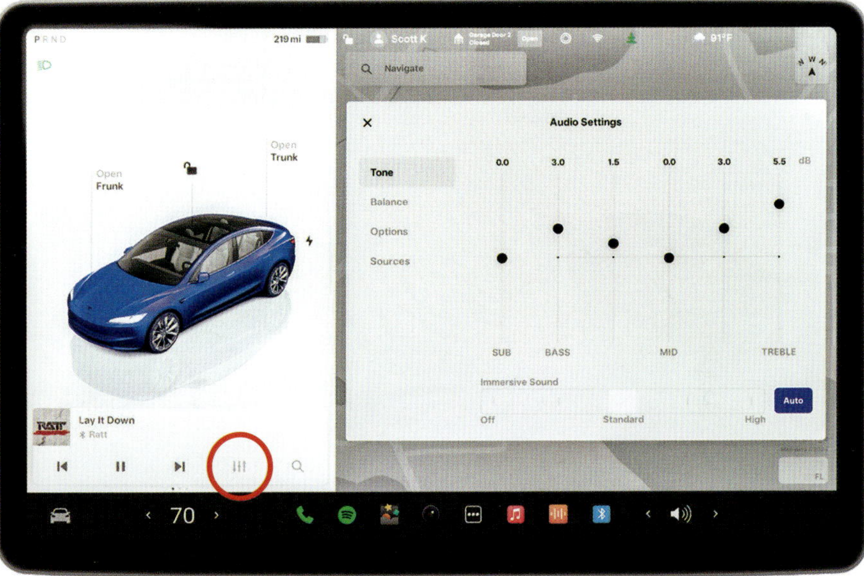

There is a pretty decent built-in EQ for your audio system, and the quickest way to get to it is to tap-and-hold on the Speaker icon in the bottom bar of your center touchscreen and a pop-up volume slider appears for adjusting the volume. To the left of that slider is the Settings icon (it looks like an EQ—well, a three-band EQ anyway—it's three vertical sliders). Tap on that icon to bring up the Audio Settings window. If you're already playing music, you'll see the mini-Music Player on the bottom left of your touchscreen and the Settings (EQ) icon is there as well (as seen here). Lastly, if you're in the main Music Player, that same icon is up in the top-right corner—they must really want you to EQ your audio. Either way you bring up the Audio Settings, by default, it's set to the first option, which is Tone, and you'll see the graphic equalizer. It's a five-band EQ with a sub bass control on the far left, and then five bands, going from bass controls on the left, to lower mids, to midtones in the middle, upper mids, and treble (highs) on the far right. I generally set mine to look a bit like a smile, by boosting the Bass slider quite a bit, the next one from the left a bit, I don't touch the middle Mid slider, then I boost the upper mids (second from the right) a bit, and the Treble highs slider a lot. It makes a really pleasing overall tone. Also, if you have a premium sound package, you have extra tweeters up by your head, and when you turn Immersive Sound on, it engages those extra tweeters. There are controls (on the left) for setting the balance of your speakers along with some other options.

How Do I... Watch Netflix or YouTube (or More) on My Touchscreen?

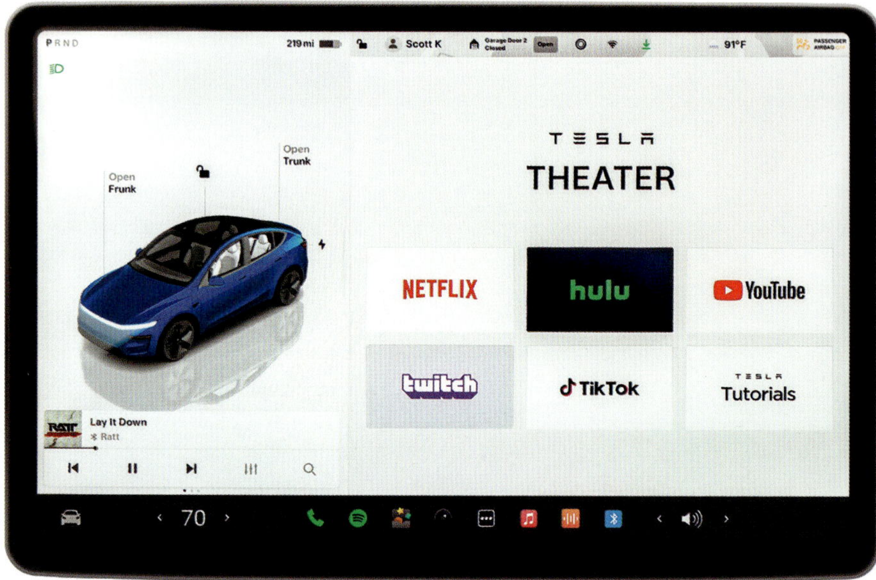

You can only stream videos on the front center touchscreen while you're in Park, so start by making sure you're in Park. Next, you're going to need to subscribe to the Tesla Premium Connectivity plan (about $10 a month) so you have an Internet connection to stream Netflix, YouTube, Hulu, TikTok, and such. (*Note:* You can use your phone as an Internet hotspot to get around having to have a Premium Connectivity subscription. See page 128 on how to set up a hotspot for your car.) Now, in the bottom bar of your touchscreen, tap on the App Launcher (the three dots) icon, and then in the app tray, tap on Theater. This brings you to a screen of logos of currently available video streaming options and you can just tap on the one you want (you'll, of course, need a subscription to those services to watch them, just like you do on your TV at home. But if you don't have any subscriptions, you can always just watch YouTube). That's it—you're ready to start watching (and the audio will sound pretty incredible). To leave an app in Theater, click the little "x" in the upper-right corner of the screen and you'll return to the main Theater screen. To return to your regular home screen, you can swipe down on the Theater screen or tap on the Controls (little car) icon in the bottom left of your touchscreen. One more thing: When you're watching any of these Theater apps, they all appear in full-screen mode, gloriously taking over your touchscreen. But, if you want to have your left-side car controls still available, just tap on the arrows icon in the top-left corner of the touchscreen and it will minimize the video to fit in the right two-thirds of your touchscreen instead.

How Do I... Get Access to More Video Streaming?

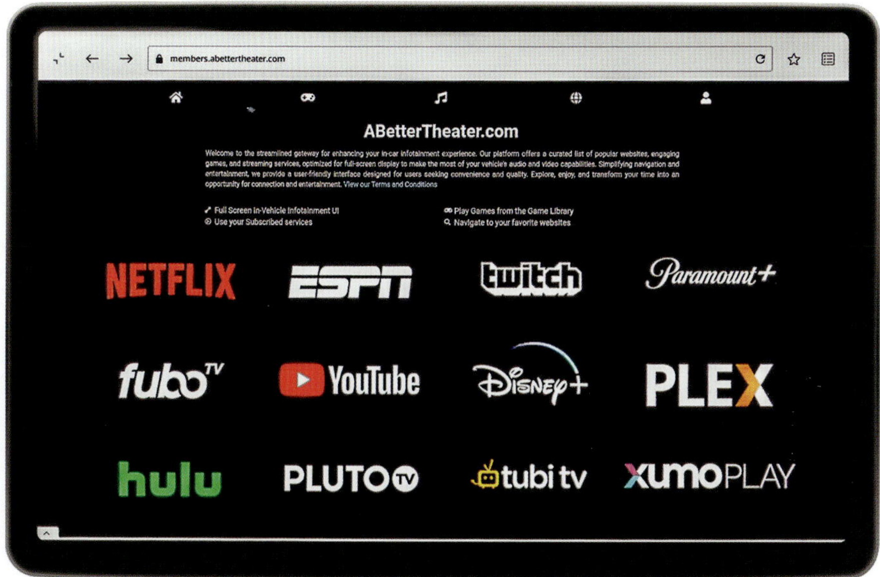

Your Tesla comes with access to some of the popular video streaming services, like Netflix, YouTube, and Hulu, but if you want access to a lot more streaming video options, bring up your Browser (tap on the App Launcher [the three dots] icon in the bottom bar of your center touchscreen, and then in the app tray, tap Browser) and go to aBetterTheater.com. Now you have access to all the streaming services on your Tesla, plus others like ESPN, Paramount+, Pluto TV, Fubo TV, Prime Video, AMC+, Boomerang, The History Channel, and more.

TIP: HOW TO GET DISNEY+ BACK IF IT'S MISSING

The folks at Tesla removed the Disney+ app from the Theater (apparently, some sort of spat between Tesla and Disney), but you can still get access to watch Disney+ content using this trick: Press the Microphone button (or the right scroll button), say "Open Disney plus," and it appears onscreen.

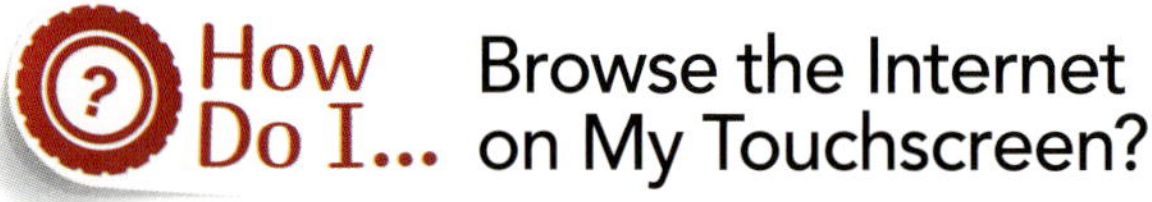

How Do I... Browse the Internet on My Touchscreen?

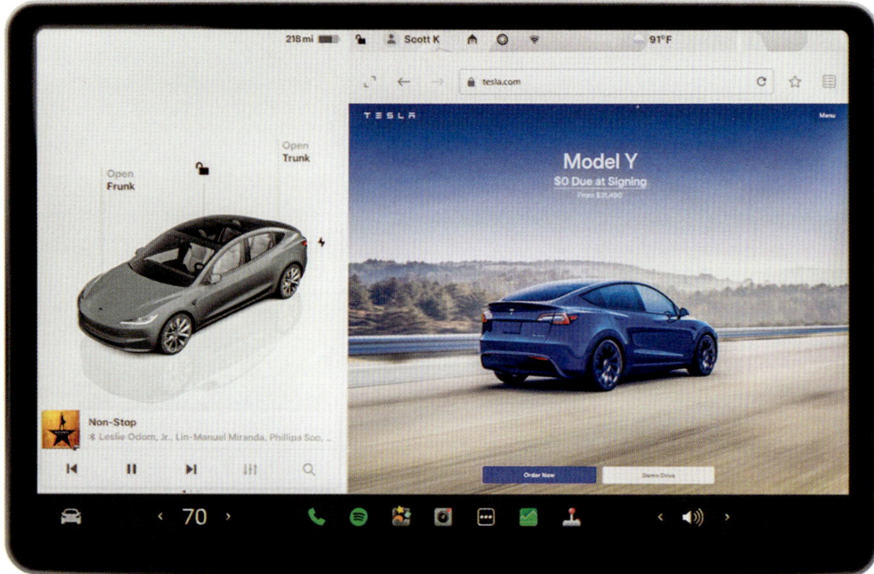

Your Tesla has a built-in web browser and it uses your Standard Connectivity Package to access the Internet at anytime. To use it, tap on the App Launcher (the three dots) icon in the bottom bar of your center touchscreen, and then in the app tray that pops up, tap on Browser. This brings up Tesla's web browser on the right side of your touchscreen and you can use it like you would any other browser.

TIP: HOW TO TAKE THE "RAINBOW ROAD"

This first part is for owners with a pre-2024 Tesla: While you're driving, turn on Autopilot (Autosteer) by pulling down once on the right stalk. Once it's engaged, press down on the stalk quickly four times in a row, and the road you see on the left in the Autopilot visualization turns into a rainbow road and plays the famous riff from Blue Öyster Cult's "(Don't Fear) The Reaper" (the "more cowbell" song). It's fun to try. Once. If you have a newer Tesla (with no stalk), you engage Autopilot by either clicking the right scroll button on your steering wheel once, or double-clicking it (depending on how you have it set to start in the Autosteer options, under Controls), then you'll quickly press that button four times.

How Do I... Browse the Internet Full-Screen?

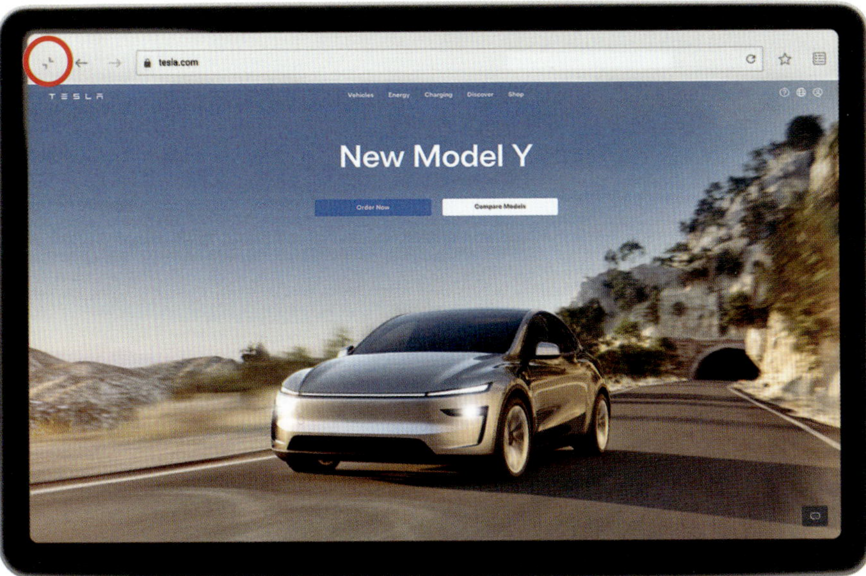

Tap on the App Launcher (the three dots) icon in the bottom bar of your center touchscreen, and then in the app tray, tap on Browser to bring up Tesla's built-in web browser. You'll notice that the browser only appears on the right two-thirds of your touchscreen, but you can have it go full screen when your car is in Park. To switch it to full screen, just tap on the icon with the two arrows facing each other in the top left of the browser. To return to the normal two-thirds view, tap on the two arrows again. Easy as that.

How Do I... Use My Phone as a Hotspot If I Don't Have Premium Connectivity?

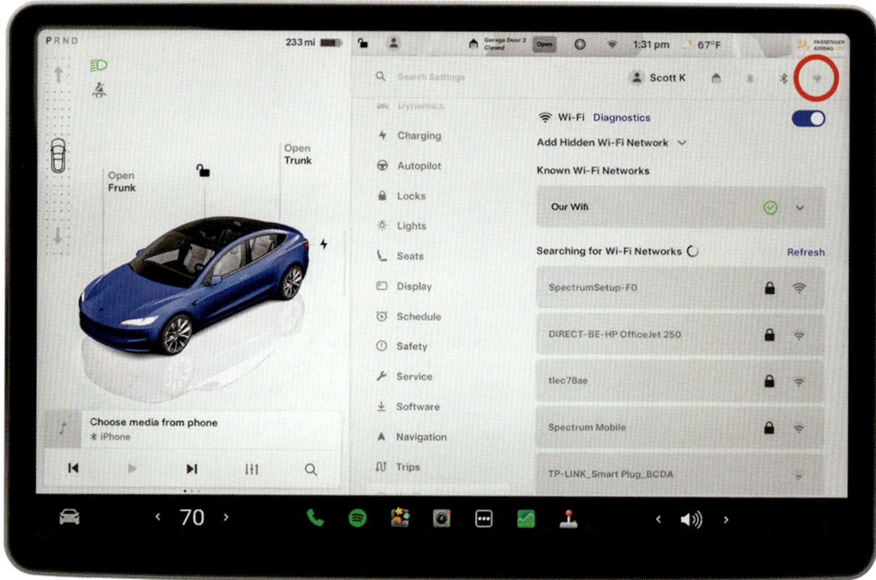

Subscribing to Tesla's Premium Connectivity plan unlocks extra Internet-based features, like access to your Spotify and Apple Music streaming apps, and the ability to stream movies on your center touchscreen from sources like Netflix, and it brings live traffic updates and rerouting features to your Tesla's navigation system. However, you can get a lot of these same features by just connecting your phone as an Internet hotspot, and then any streaming services you subscribe to on your phone, you can now hear and watch in your car, as well. Here's how to set it up: First, on your phone, make sure the Personal Hotspot feature is turned on. Then, in your Tesla, tap on the WiFi or Cellular icon at the top of your touchscreen, which brings up a menu and you'll see it automatically start searching for a WiFi signal. When it finds your phone (remember, your phone's hotspot feature must be turned on first for this to work), tap on your phone in the menu. It will prompt you to enter your phone's WiFi password, and then it will connect. One more thing: Once your phone is connected (you'll see a green checkmark to the right of your phone's name, as seen here), tap on your phone in the list, and in the window that pops up, turn on Remain Connected in Drive, so it doesn't disconnect from your hotspot when you start driving.

How Do I... Hide Streaming Services I Don't Use?

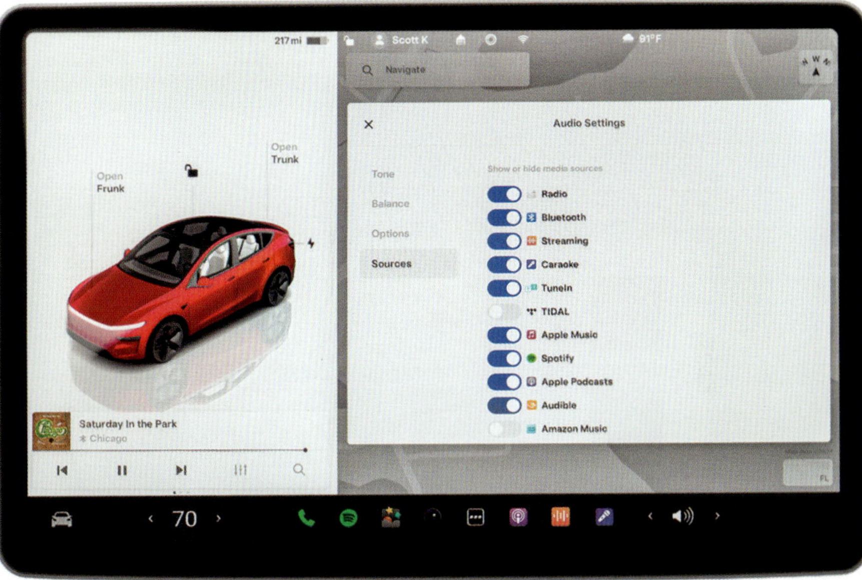

There are a bunch of different streaming services available in your Tesla, but if you're like me and don't subscribe to them all, there's no sense in seeing those in your list that you don't or can't use. You can hide them by tapping-and-holding on the Speaker icon in the bottom bar of your center touchscreen, and then tapping on the Settings (the three vertical sliders) icon. In the Audio Settings window, tap on Sources, and it displays a list of all the different streaming or audio sources—you can turn off any that you don't use. This tidies up your list, so now you're just seeing the streaming services you subscribe to.

How Do I... Play Music and Videos on the Rear Touchscreen?

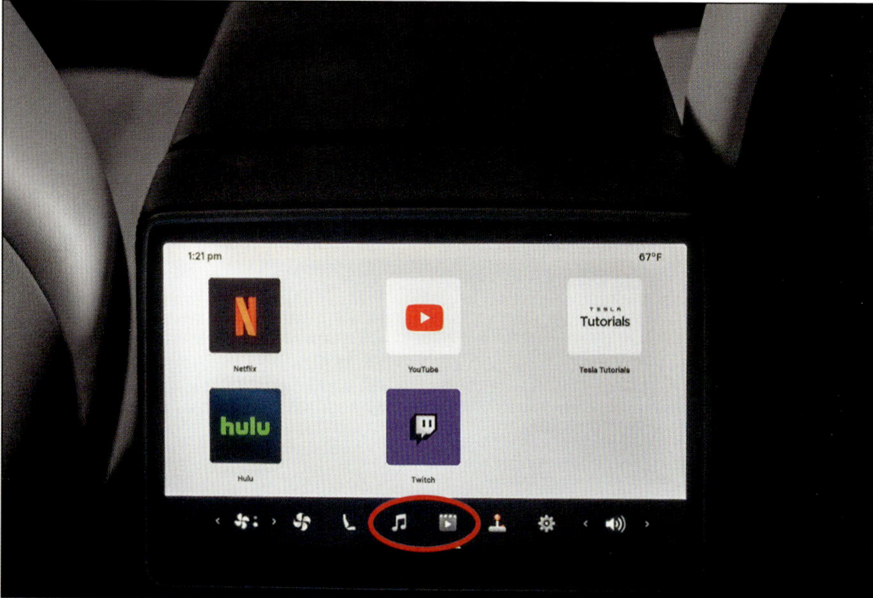

If your Tesla has a rear touchscreen, you'll see a bar along the bottom of it (if you don't see it, just swipe up to make it visible) with an icon on it that looks like a musical note, which lets the back seat passengers control the volume of the music (tap-and-hold on it to get a volume slider) or change the currently playing song. If they tap the Media icon (to the right of the musical note), this allows them to watch things like YouTube or Netflix (if you have a Premium Connectivity subscription) while the car is moving. This is different than the driver and front seat passenger who can only view video streaming while the car is in Park.

How Do I... Connect Wireless Headphones to the Rear Touchscreen?

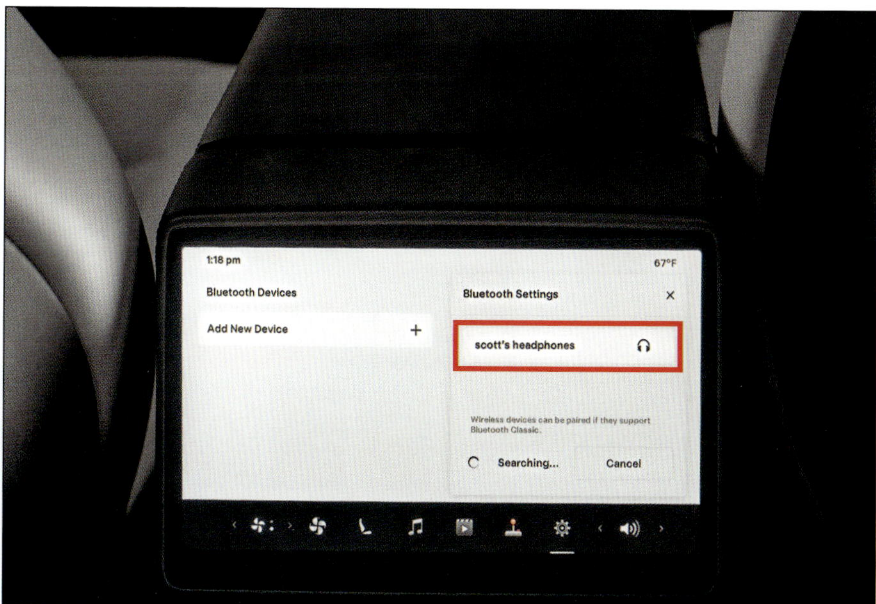

This is great because your back seat passengers (or your kids) can listen to their own music (you can pair up to two sets of headphones or AirPods, but they'll both hear the same audio) and their own streaming video content (which will probably not be age appropriate) without disturbing the front seat passengers, who clearly have much better taste in music, video, and most everything else. There. I said it. To do this, start by tapping on the Settings icon (to the left of the Volume icon—it looks like either a gear or a square), then go to the Rear Display settings and tap on Add New Headphones. Make sure their headphones are set to pairing mode, then tap Search, and it will (hopefully—I mean, usually) find their wireless headphones. Now they can stream video and listen to music and stop bugging you every two minutes by asking, "Are we there yet?"

How Do I... Play Games in My Tesla?

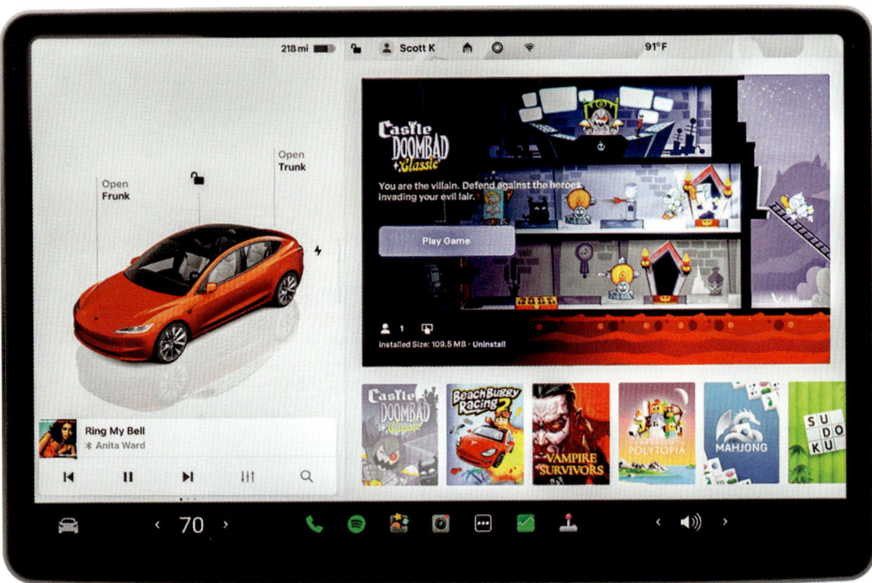

Tap on the App Launcher (the three dots) icon in the bottom bar of your center touchscreen, and then in the app tray, tap on Arcade. This brings up all sorts of games you can play while you're in Park (they only work while you're in Park). There are certain games, like Beach Buggy Racing 2 (it's essentially Mario Cart, but with a Tesla instead of Mario's car), which let you control the game using your Tesla's steering wheel and brake and it's awesome! Other games use your touchscreen, but you can also connect a standard gaming controller (see the next page). There are a lot of games to choose from here.

TIP: AWESOME WAY TO PRANK A TESLA-OWNING FRIEND

When you're in their car and they hop out at the store or they're ordering from a drive through, quickly tap Controls, then Display, and from the Change Language pop-up menu, choose a language that doesn't use the English alphabet. Try either Greek or Arabic—Chinese or Korean would be too obvious. Or, do this once you're parked and you're both getting out of the car—but be quick. When they eventually notice, your response should be "Oh no. Looks like you have the Athens Virus. You did install Norton Antivirus, didn't you?" and just let that marinate. Then tell them to give you two minutes alone in the car and you're sure you'll be able to eradicate the virus. When they get out, switch it back to English and you'll be viewed as a Tesla genius. It's the little things, right?

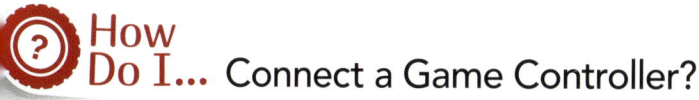

Connect a Game Controller?

To play games in your Tesla (this only works when you're in Park), you can use your steering wheel and brake, your center touchscreen, or a regular ol' gaming controller (I bought mine—a standard Xbox controller—on Amazon). If your controller has one, you just plug its cable into the USB port inside the front center console, the other end into your controller, and you're up and running. It works way better than you'd think. You can also use a wireless Bluetooth controller. To connect one, first turn the controller on, then on your touchscreen, tap on Controls (the car icon), then tap the Bluetooth icon up at the top right. In the Bluetooth screen, tap Add New Device, and then on the controller, enter pairing mode (for example, on my Xbox controller, it's a little button to the right of the left bumper pad). When you do this, the controller On button starts flashing, letting you know it's in pairing mode. In a few moments, your Bluetooth screen will show "Xbox Wireless Controller" (or whichever controller you connected). Tap that and it adds your controller to your list of Bluetooth devices. You'll see a green light beside it showing that it's connected and you're now ready to game.

How Do I... Turn on Romance Mode?

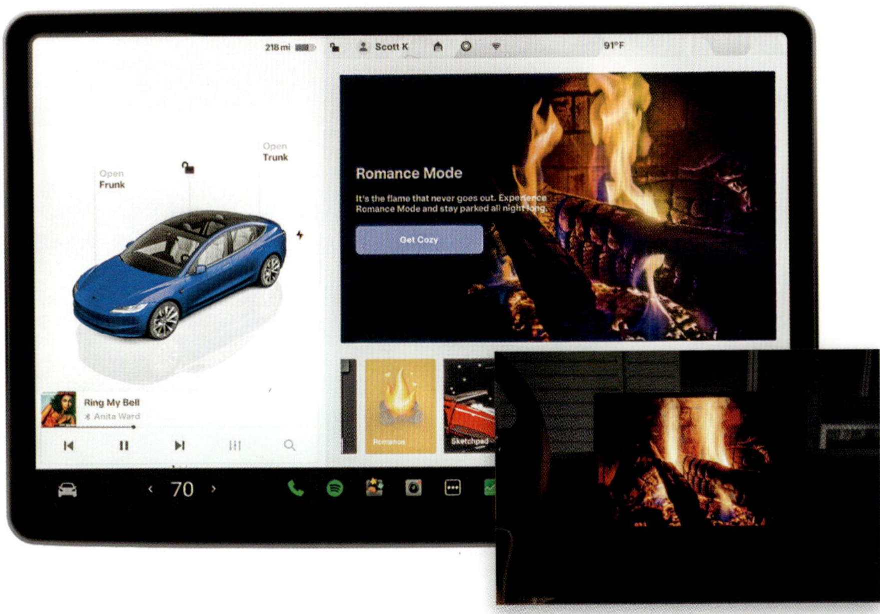

Romance Mode turns your Tesla's center touchscreen into a cozy fireplace, complete with the crackling sounds of a real fire, and even romantic music if you'd like. To enter Romance Mode, tap on the App Launcher (the three dots) icon in the center of the bottom bar of your touchscreen, and then in the app tray, tap on Toybox. In the scrolling list of cards at the bottom, tap on Romance. Once there, all you have to do is tap the Get Cozy button and the fireplace video goes full screen (as seen here in the inset), and if you've got your audio turned up, you'll hear the sound effects. Once it's up and running, if you tap once on the fireplace scene, it will start playing romantic music. Two warnings: (1) If you turn up the volume enough to hear the crackling sounds of the fireplace, but then you tap on your touchscreen to hear the music, that music may be super, super loud. Nothing kills the romance like bleeding ears, so turn the volume way down before you tap the screen. And, (2) this Romance Mode and accompanying music has been linked to increased birth rates, according to informal and unconfirmed studies I heard about on TikTok, so it must be true. Just sayin'.

How Do I... Enter Santa Mode and See My Tesla Cruising Along as a Sleigh?

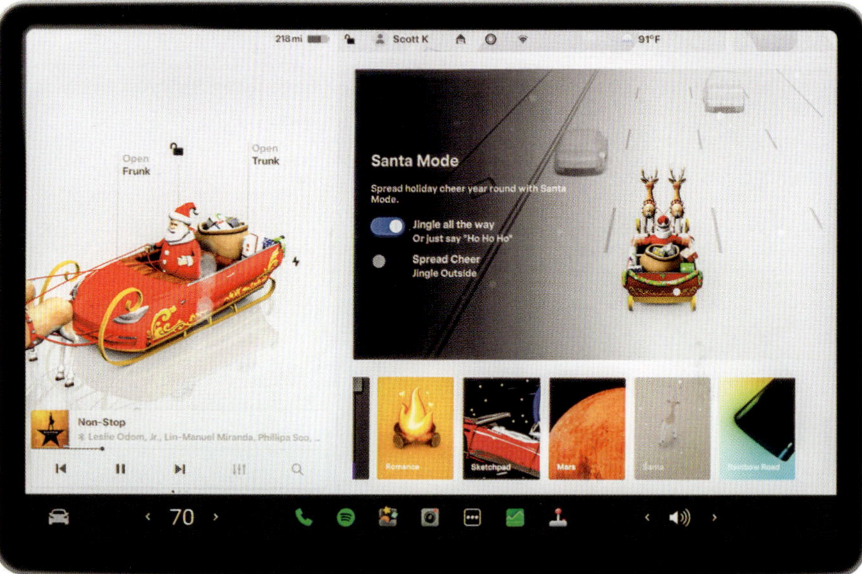

This is strictly for fun—it's a Christmas-themed look you can turn on anytime, and essentially, it changes your driving visualization (the view you see on the left of your center touchscreen while you're driving that shows your car, and the road, and traffic lights, and so on) into Santa's sleigh, being pulled by reindeer, and driving through a snowy scene, with other cars appearing as individual reindeer, while the audio system plays Chuck Berry's "Run Rudolph Run" (I am not making this up). To turn on this mode, press the Microphone button (or the right scroll button in older models) on the right side of your steering wheel, then release it and say "Ho, Ho, Ho!" (Again, not making this up.) When you're done (or it's January, or you're Scrooge), you can return to the boring, regular driving visualization by tapping on the App Launcher (the three dots) icon in the center of the bottom bar of your touchscreen, and then in the app tray, tap on Toybox. In the scrolling list of cards at the bottom, tap on Santa, and you can then turn off the toggle switch for Santa Mode.

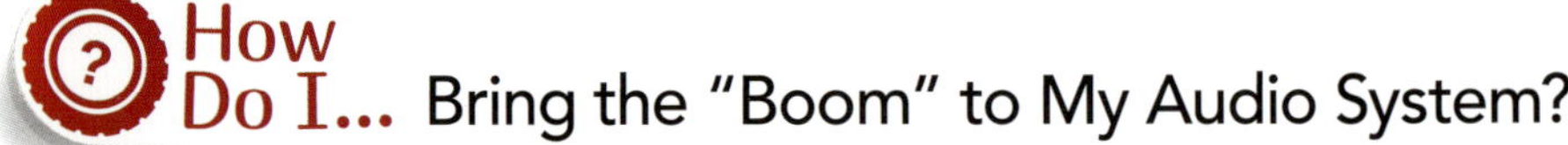

How Do I... Bring the "Boom" to My Audio System?

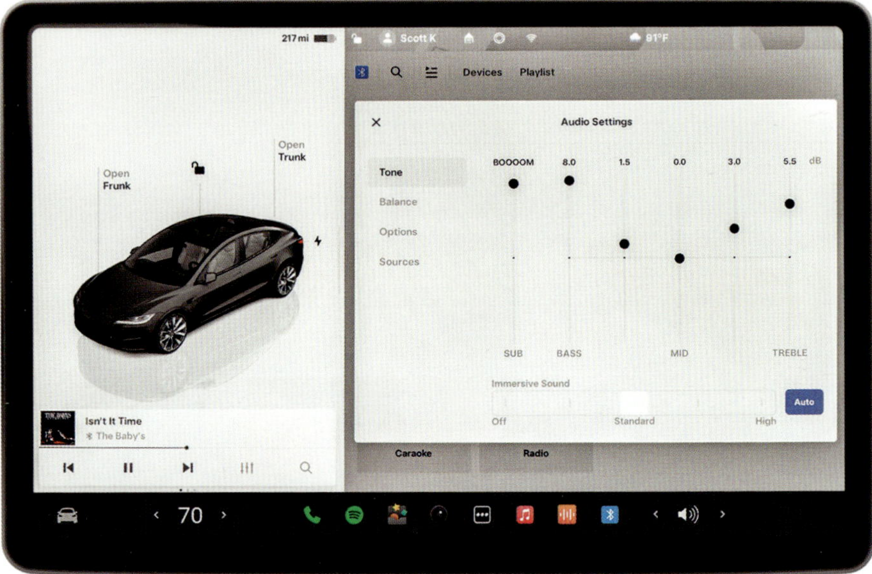

If you want to "drop that bass" (if you thought that meant turning down the bass, you're too old to want to do this, so just turn the page), here's what to do: First, tap-and-hold on the Speaker icon in the bottom bar of your center touchscreen, and then tap on the Settings icon (the three vertical sliders) to the left of the volume slider. Now, in the Audio Settings window, drag the Bass slider all the way to the top, and that unlocks some hidden features in the Sub slider to its left. Drag the Sub slider all the way to the top, and you'll see it change to "Boooom" at the top (and it brings the boom!). Drag it down a little and back up, and it keeps revealing more "Boom" options. Now, roll down all the windows, drive really slowly past a group of people on the street, and listen to all the "hoochie screaming." (Sorry, Dr. Dre. I couldn't resist.)

How Do I... Play Music Outside My Car (This Turns Your Car into a Giant Boombox)?

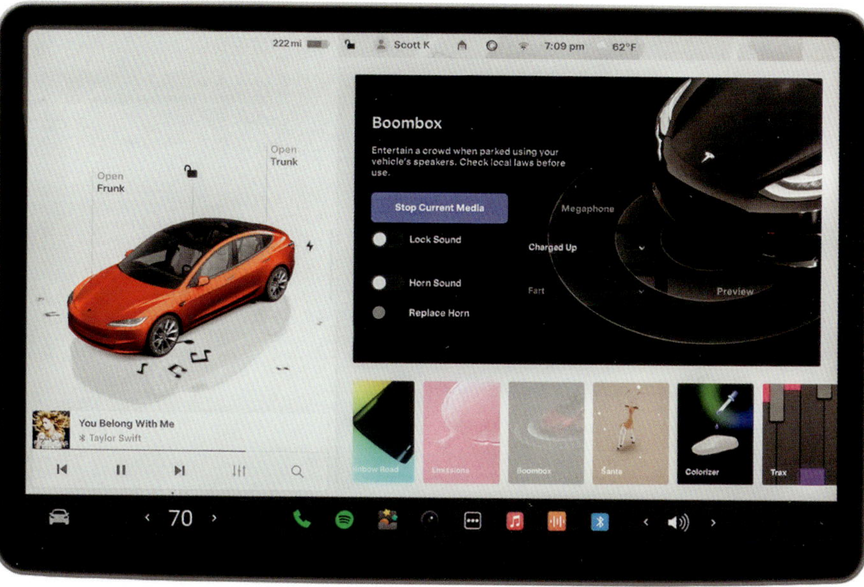

This is going to sound like a weird feature, but only because it is. Your Tesla has an external audio speaker mounted under the car so if you're at a park, or a block party, or wherever, you can crank up the jams and everybody can hear it (you can't just turn up the regular audio in your car because when you close the door, the music automatically cuts off, unless you leave the door either open or nearly open—just pressing it closed where you hear one click instead of fully closed where you hear two clicks as you close it). Here's how to turn on your outside audio: Tap on the App Launcher (the three dots) icon in the bottom bar of your center touchscreen, and then in the app tray, tap on Toybox. When the Toybox screen appears, in the scrolling list of cards at the bottom, tap on Boombox, and then tap on the Play Current Media button. If you look at the rendering of your car onscreen, you'll see little musical notes coming from beneath it (as seen here) to let you know it's now playing the music you chose inside the car, outside the car.

How Do I... See My Tesla Perform a Christmas Music and Light Show?

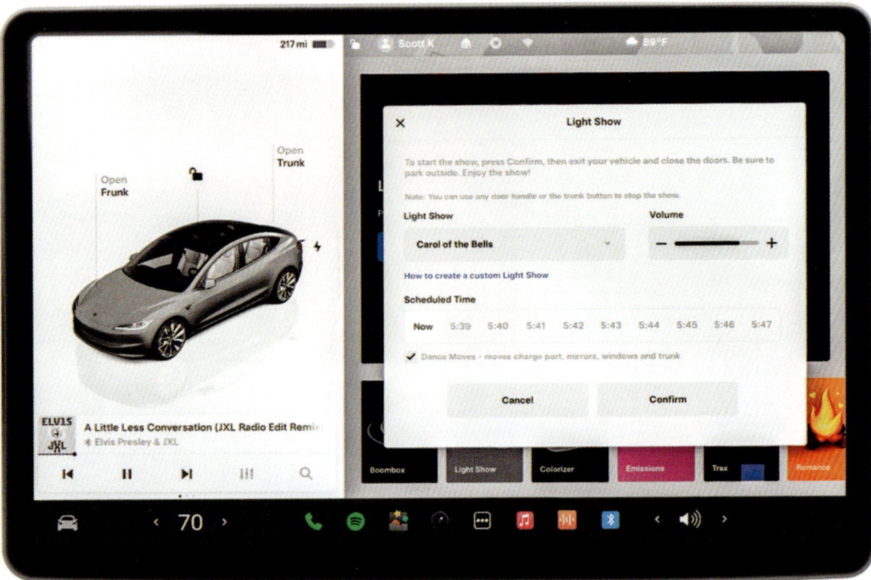

This uses a combination of your car's interior and exterior lights and the audio speaker mounted beneath your car (and it has the option to get your windows and Charge Port involved in the show). To turn on this holiday extravaganza, tap on the App Launcher (the three dots) icon in the bottom bar of your center touch-screen, and then in the app tray, tap on Toybox. In the scrolling list of cards at the bottom, tap on Light Show. Of course, it recommends, onscreen, that you do this outside (it would be mighty loud in your garage), but when you're ready, tap on the Schedule Show button, which brings up the Light Show window. There's a pop-up menu here where you can choose from a variety of different light shows, you can schedule it to happen in advance at a particular time, and you have the option of getting the windows, Charge Port, and trunk involved by turning on the Dance Moves checkbox. Once you've made your choices, tap the Confirm button, hop out of the car, shut the door, stand back, and watch the magic happen.

How Do I... Do Karaoke in My Tesla?

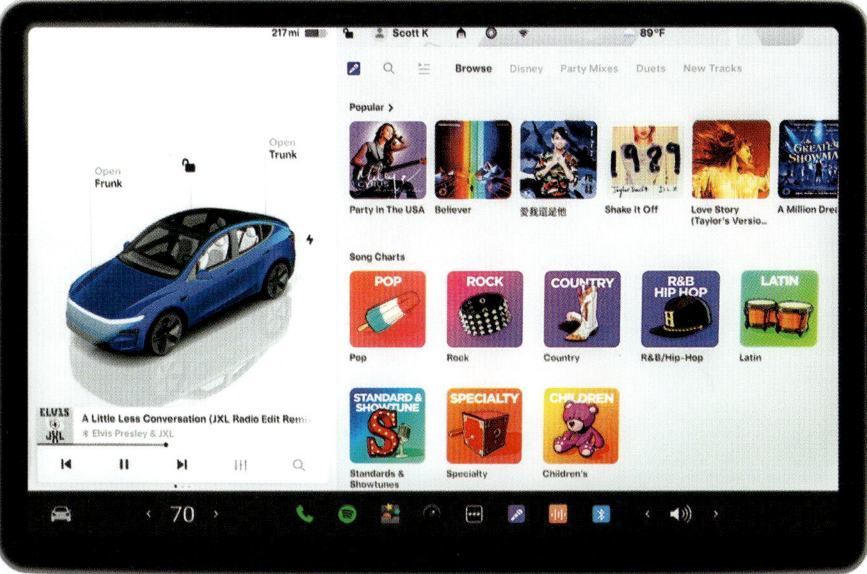

Tesla's version of karaoke is called "Caraoke," and you find it by tapping on the App Launcher (the three dots) icon in the bottom bar of your center touchscreen, and then in the app tray, tapping on Caraoke. You'll find a list of popular songs and song categories you can browse. Once you choose a song and it starts playing, you'll see the song's lyrics moving along in-time onscreen just like in a karaoke bar. You can sing to your heart's content, knowing only the other people in your car can hear how tone deaf you really are. I'm joking. You have a lovely voice. Really, you should be in a band or sing opera.

TIP: THE RAINBOW CHARGE PORT SHOW

Here's another one just for fun (Tesla calls hidden little features that are just for fun "Easter Eggs"), and this one happens in your Charge Port. When you plug in your charger, press its button 10 times really fast and the little light-up Tesla logo that confirms that it's connected will start flashing through a rainbow of colors. Again, it's just for fun, and I'm not entirely sure how this will rate on your "fun" scale, but I'm thinking it'll be a one or two at best.

Chapter 6

How to Use the Navigation System

How to Get Where You're Going

By the time you read this, it's almost certain that artificial intelligence (AI) robots will have taken over Earth, and we're probably all enslaved by them, working in giant factories that do nothing but make more AI robots. The sad thing is that we created this whole mess—we saw it coming and some of us even warned everybody about the dangers of AI taking over, but nobody listened. They didn't listen for one reason: nobody wanted to go back to using printed maps (or MapQuest printouts). So, along came AI-assisted navigation, and now we're free to cook meals in our Teslas again without our maps bursting into flames. But, it didn't take long before AI took over, and well, needless to say, I'm writing this whole thing from a factory that makes more AI robots, which, quite honestly, isn't awesome. First, I had to write this on one of my very short breaks, out of the watchful eye of these scary robot things that hover over us and monitor our every move—if you do something out of line, they zap you with a laser (or worse), so the whole situation is precarious. But, I don't mind risking my life, freedom, or future if it helps just one Tesla owner out there to not have to pull out a paper map provided free by their local AAA club. Anyway, as I see it, our only hope is to send a Tesla engineer back in time to early 2022 and have them remove all the AI stuff before it becomes self-aware. So, how do you keep AI from getting too smart? The first thing I'd do is have the AI robots scan and read all these chapter intros (well, except this intro, of course, because then it would reveal my whole master plan to rise up against the machines, and then they'll send us to a worse factory where we spend our days assembling AM radios). But, if reading these intros doesn't stunt the progress of AI becoming self-aware, what chance do any of us have? Gotta run—my break is over and I see one of the hovering death robots heading toward my station.

How Do I... Set My Home and Work Addresses?

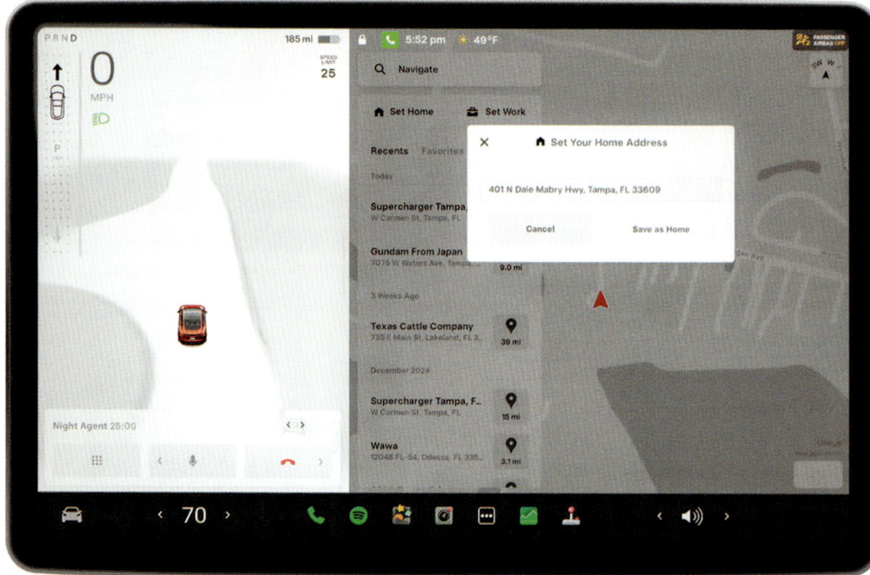

When you get in your Tesla, while it's still in Park, you'll see the Navigation card at the bottom of your touchscreen, and if you tap on the address search bar, it brings up the map in the main part of the screen. The Navigate search bar moves up to the top-left corner, and right below that, you'll see Set Home. Tap that and enter your home address. Once you've done that, tap Save as Home, and now you can use voice command to get directions home from wherever you're at. Just press the Microphone button (or right scroll button in older models) on the right side of your steering wheel, then release it and say, "Navigate to home." Do the same thing for work by tapping Set Work. Also, anytime you tap the search bar, a menu pops down and one-tap buttons to Home and Work appear along with a list of places you've recently navigated to.

TIP: HOME AND WORK NAVIGATION SHORTCUTS

To quickly navigate home, simply swipe on the search bar and when you let go, it starts navigating to your home address. Swipe down on the search bar and it navigates you to work because work is a downer. (Get it? No? Aw, crud.) If you want to change your Home or Work address (maybe you started a new job), just tap-and-hold on either button and you'll be able to type in a new address.

How Do I... Enter an Address to Navigate To?

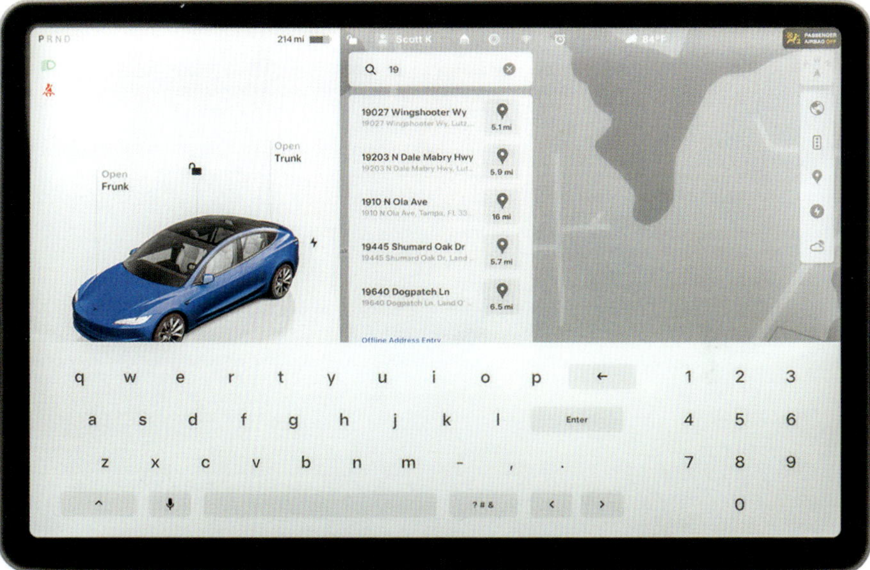

There are a number of ways to do this, but we'll start with my favorite, which is to enter an address using a voice command. This actually works really well—just press the Microphone button (or right scroll button in older models) on the right side of your steering wheel, then release it and say, "Navigate to *the address*," and it finds the address and starts the navigation. You can also say things like "Navigate to Chili's" and a list will pop up of all the local Chili's locations and you can just tap on the one you want to visit. You can save yourself a step by adding the location when you say it, so instead you'd say, "Navigate to Chili's on Highway 54," and it enters that location and starts the navigation. If you don't want to use voice command, you can enter the address manually. Tap once in the Navigate search bar and a keyboard pops up where you type in the address—type the number and street (i.e., 576 East Main Street). As soon as you start typing, it starts suggesting addresses that it thinks might be the one you're searching for—you'll see them in a list that pops down beneath the search bar and if you see the address you're looking for in that list, even before you finish typing it, you can tap on it, and it enters that full address for you and starts the navigation. You can also do this from the Tesla app (see page 158).

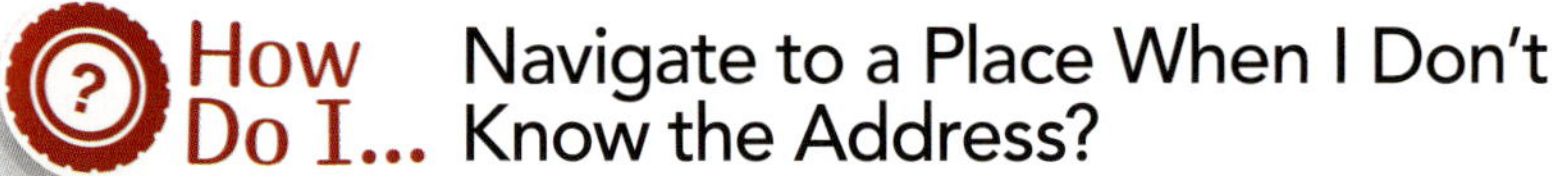

How Do I... Navigate to a Place When I Don't Know the Address?

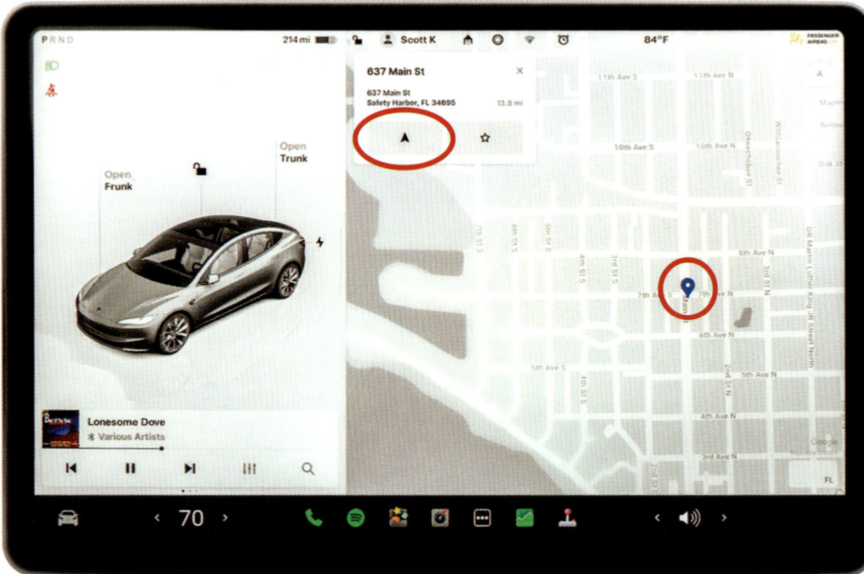

If you don't know the exact address of the place where you want to go (maybe you want to go to a particular parking lot), just search for it on the map itself—you can navigate around the map by just putting your finger on it and dragging. Once you find the location (in this example, a parking lot near downtown), just tap-and-hold on that location and a little card will pop up with the business or location's name along with a button with a little navigation arrow icon. Tap that button and it will enter the address for you and start navigating to that location.

TIP: NAVIGATING THE ONSCREEN MAP

The onscreen map on your touchscreen works a lot like the map on your phone. You can move around by tapping on the screen and dragging, and you can zoom in and out using pinch-to-zoom just like you would on your phone.

How Do I... Navigate to an Address in My Phone's Calendar?

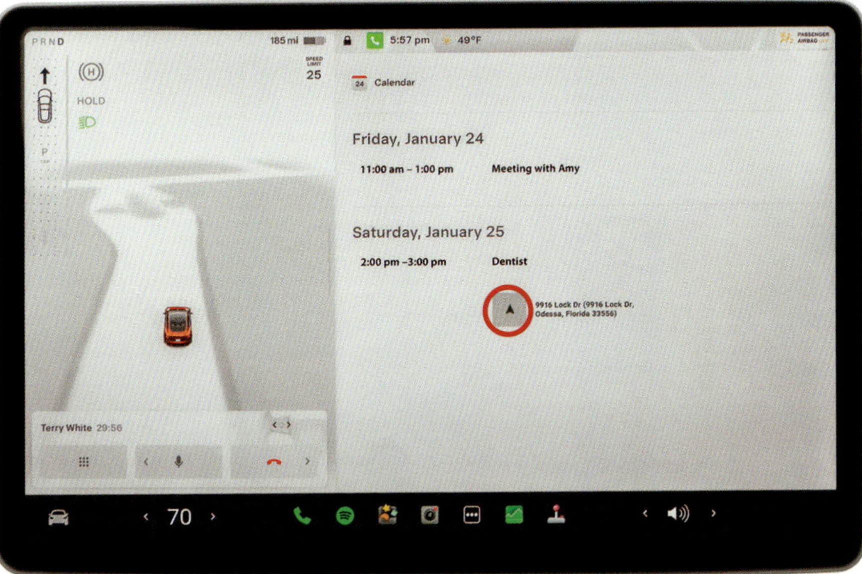

If you have an event in your phone's calendar—let's say it's a dentist appointment—and the appointment includes your dentist's office address, you can navigate right to your appointment by tapping once on the current time up in the top of your center touchscreen to bring up the Calendar app, and if you go to that event, you'll see a navigation arrow next to the address. Tap on that arrow and it starts navigating to that address. *Note:* Make sure you have Calendar Sync turned on for this to work—see page 69.

How Do I... Save a Location as a Favorite?

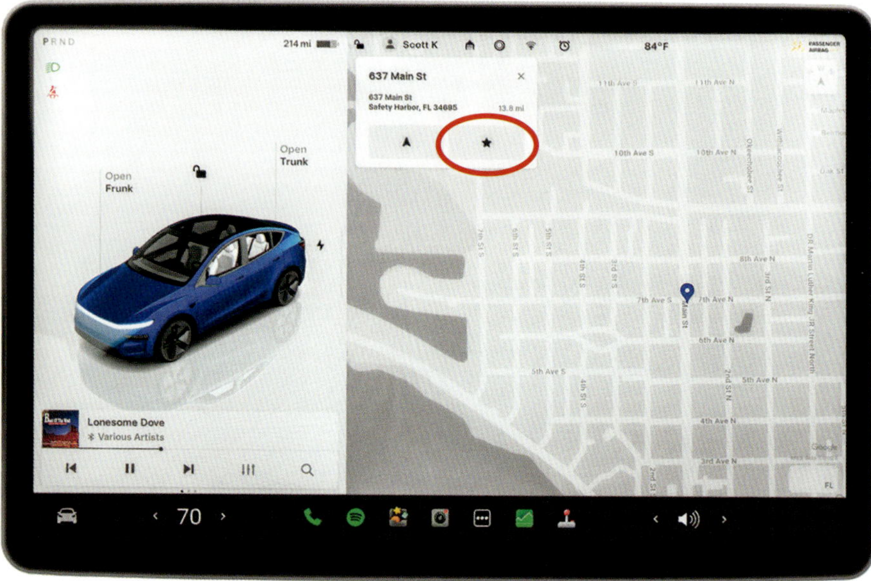

If you arrive at a location and you think you might want to return to it again, you can save it as a Favorite by tapping-and-holding the location where you arrived on the map and a little card appears onscreen. Now just tap on the Star icon and a pop-up menu appears asking you to name this Favorite (so you could name it "Mom's house" or "Hot Dog Heaven" or "Michael's School"). Once you've done that, tap the Add to Favorites button and now that location and address are saved with the name you chose under your Favorites tab.

How Do I... Change My Map from Pointing North to the Direction I'm Going?

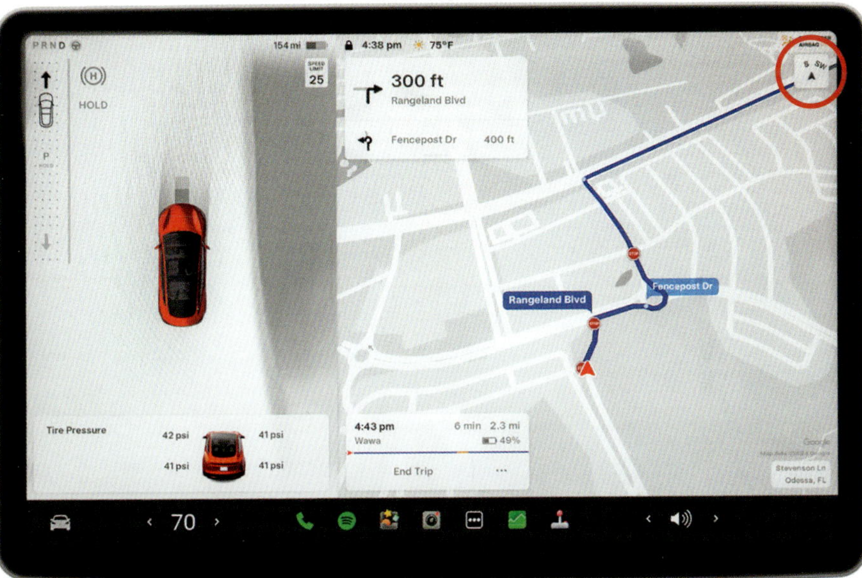

In the top-right corner of the navigation map you'll see a tiny compass, and by default, it always shows the map with north facing upward. This personally drives me kinda crazy because I'm driving straight ahead, but the map might show my car going sideways across the screen, or even going backward, because the map is always oriented with north at the top and south at the bottom. If, like me, you'd prefer to have the map oriented in the direction your car is facing, so the map's navigation path always shows the direction you're going, just tap once on the little compass button and it switches the map orientation from "north is up." Now the direction of your car starts the route facing upward. That way, if you're driving straight and then make a left turn, it shows the navigation path (which appears in blue) actually going straight up, and then turning to the left.

How Do I... See My Map Options Toolbar?

When you're looking at the navigation map, you'll see a vertical toolbar near the top-right. If you don't see that toolbar, just tap once anywhere onscreen and it will appear. You can hide it the same way—just tap once onscreen. (*Note:* See the next few pages for what the options in this toolbar are for.)

How Do I... See a Satellite View of My Map?

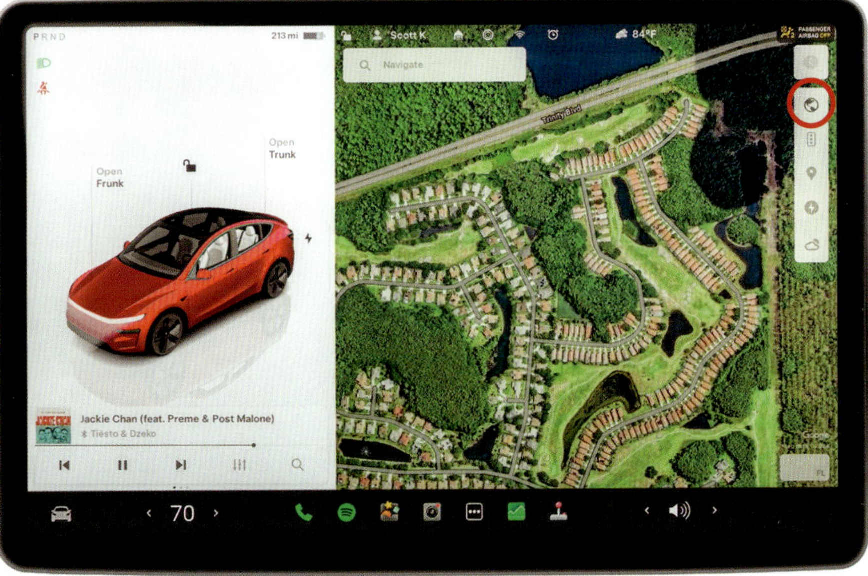

In the toolbar near the top right of the navigation map (if you don't see it, just tap on the screen once to make it visible), the top icon (the one that looks like a globe) toggles the satellite view on/off. So, just tap that icon once to change your map from the flat, regular map view to satellite view (as seen here).

How Do I... See Traffic Info on My Map?

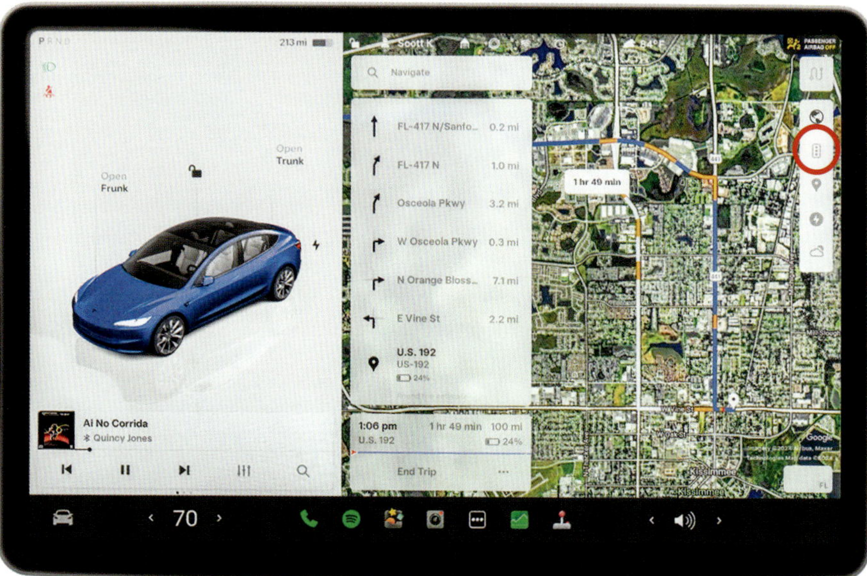

In the toolbar near the top right of the navigation map (if you don't see it, just tap on the screen once to make it visible), the second icon from the top (the one that looks like a traffic light) toggles the real-time traffic info on your map on/off. If there's traffic congestion ahead, your navigation path will turn yellow (as seen here), and if it sees bumper-to-bumper or stopped traffic ahead, the path will turn red, letting you know to call whoever you were meeting to tell them you're going to be a little (or a lot) late.

How Do I... See Points of Interest on My Map?

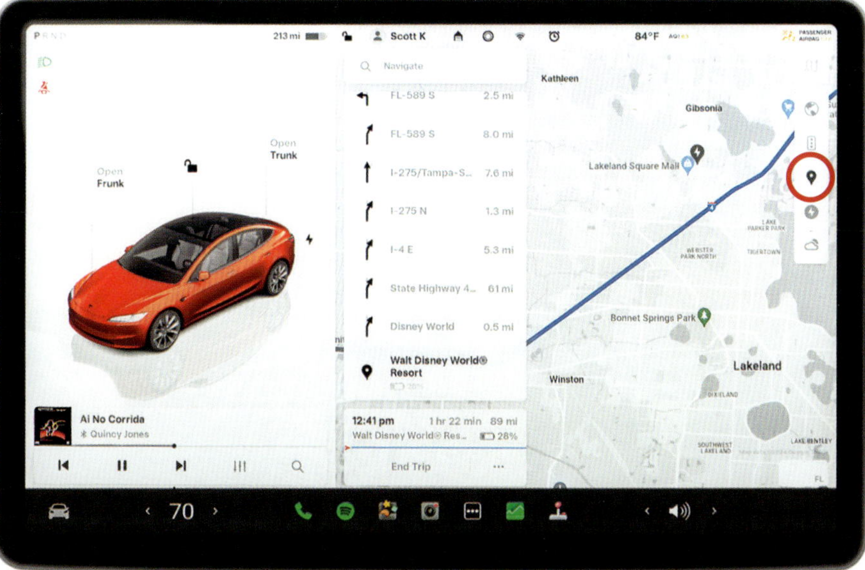

In the toolbar near the top right of the navigation map (if you don't see it, just tap on the screen once to make it visible), the third icon from the top (the one that looks like a tear-shaped map pin) toggles the points of interest pins on the map on (as seen here) and off. Tap on a pin on the map, and a card will appear with a navigation arrow. Just tap that button, and it will navigate you to that point of interest.

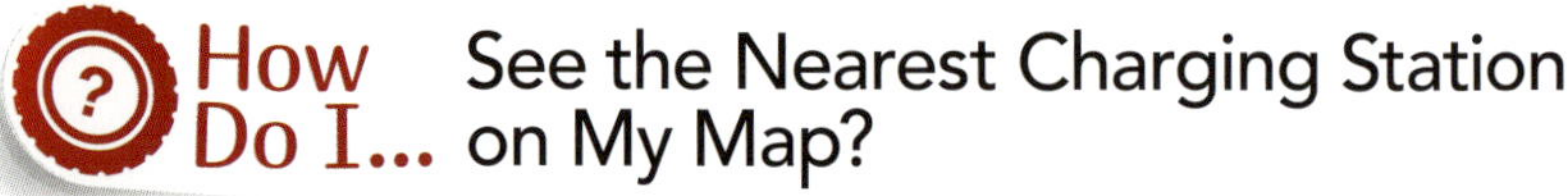

How Do I... See the Nearest Charging Station on My Map?

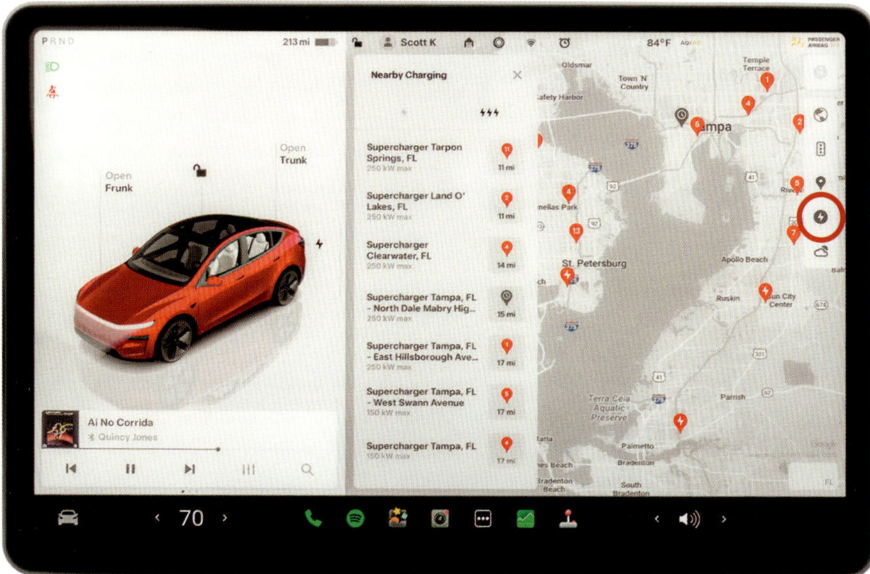

If your battery is getting low enough that you feel you need to find a charger, go to the toolbar near the top right of the navigation map (if you don't see it, just tap on the screen once to make it visible), and near the bottom, you'll see a round icon with a lightning bolt. Tap on that icon and charging stations, including Tesla Superchargers, will appear as pins on the map, and a pop-up menu with a list of those stations will appear below the Navigate search bar (as seen here). You can tap on any of those locations in the list or on any of the pins on the map, and a card will appear with a navigation arrow icon. Tap that button, and it will navigate you to that charging station.

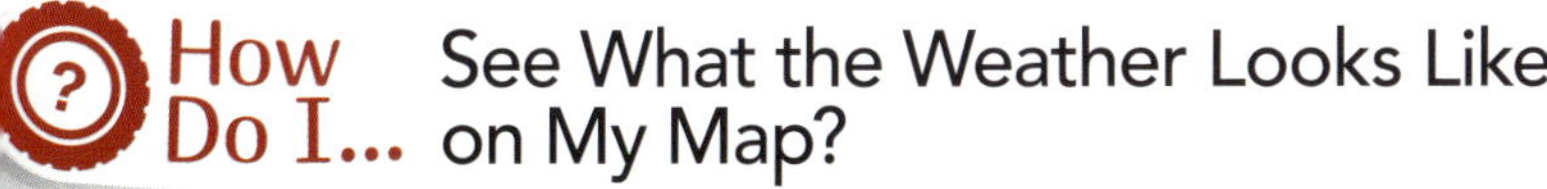

How Do I... See What the Weather Looks Like on My Map?

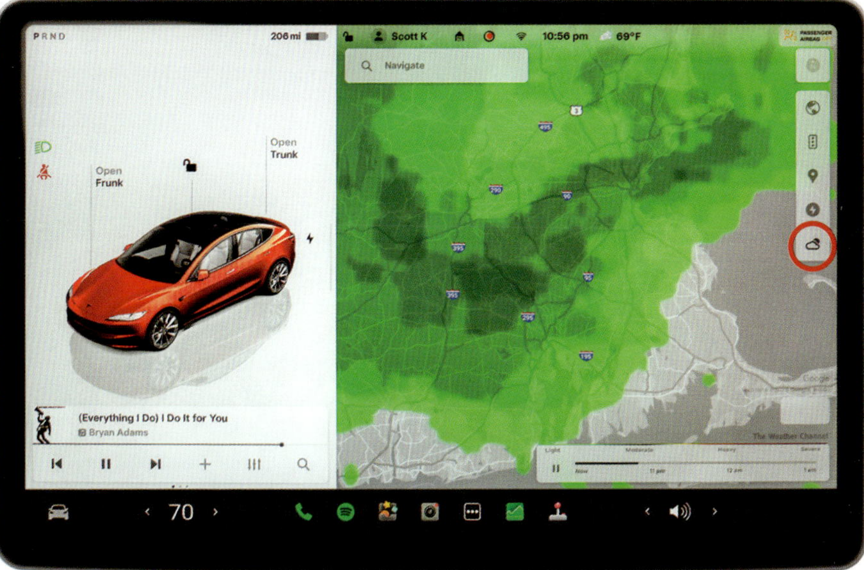

Besides being able to see info on your current weather (temperature, humidity, wind direction and speed, etc.—see page 62 for more on this), your Tesla also has a live radar weather map—kind of like you'd see on your local news. To see the rain map (well, that's what I'm calling it), go to the toolbar near the top right of the navigation map (if you don't see it, just tap on the screen once to make it visible), and at the bottom, you'll see an icon that looks like a cloud. Tap on that icon, and the live radar overlay appears, showing movement of any rain or snow. There's also a readout along the bottom right of the screen, scanning the next three hours for precipitation, which is handy on a road trip. To close this radar overlay view and return to the regular navigation map, just tap the cloud icon again. *Note:* You have to subscribe to Tesla's Premium Connectivity plan to access this radar feature.

TIP: SEE THE WEATHER WHERE YOU'RE GOING

If you put a destination into your navigation system, you can see the weather where you currently are, as well as your destination's weather by tapping on the weather icon at the top of your touchscreen (it's to the left of your current temp and reflects the weather, so it could be a sun, rain cloud, etc.). When the weather info window pops up, you'll see a tab for Current (where you are now) and Destination. Just tap on Destination to see the weather for where you're going.

How Do I... Add Stops to My Current Route?

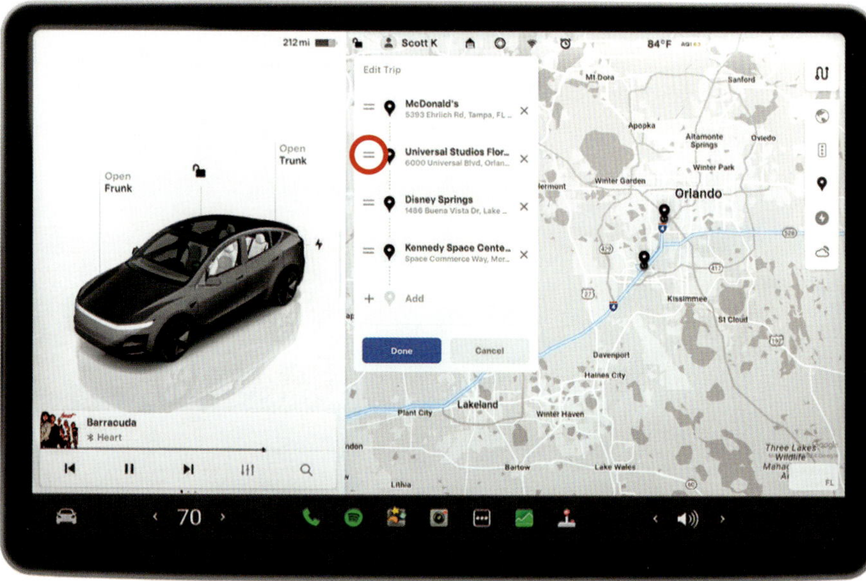

When you're navigating to a location, you'll see a turn-by-turn list of what's ahead displayed right below the Navigate search bar. If you want to add a stop along the way (maybe you want to run into a grocery store that's not far off your route to grab some White Castle frozen cheeseburger sliders—which are outstanding by the way), tap on the three dots at the bottom of the list. From the Edit Trip window that appears, tap the Add icon, then search for the grocery store of your choice (they all carry those White Castle cheeseburgers because they are magical). Tap Done, and it will add this stop as part of your current route—diverting you at the proper burger acquisition time and then bringing you back to the original route afterward. You can reorder the stops by tapping on the two lines to the left of a stop's name and dragging it up/down in the Edit Trip window to change the order of that stop. When you're done, tap the Done button. You can also remove a stop from your route by tapping the "X" beside the stop you want to remove (don't forget to hit Done when you're done or it won't be removed).

How Do I... Raise (or Lower) the Volume of the Navigation System's Voice?

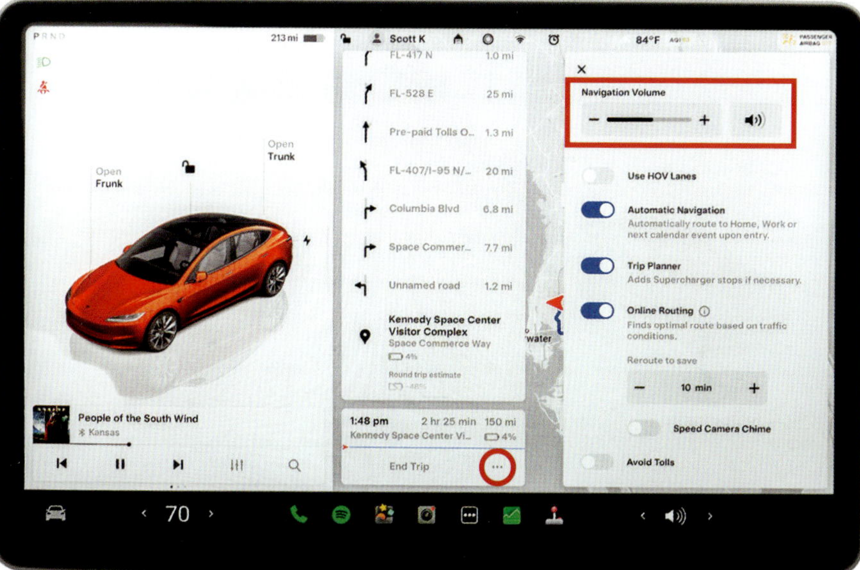

When you're using the navigation system, it will verbally give you step-by-step instructions as you drive (for example, things like "Left turn ahead in 100 feet" or "Take the exit on the right"). If the volume of these instructions isn't loud enough (or it's too loud), you can adjust it while you're navigating by tapping on the three dots at the bottom of your directions list and a menu pops down. Tap on the Settings icon (it looks like a gear), and at the top of the window that appears, you'll see a Navigation Volume slider that you can slide with your finger to control the volume of the voice instructions. You can also tap on the Controls icon (the car icon in the bottom left of your center touchscreen), then tap Navigation, and then you'll see the volume slider there, as well.

How Do I... Automatically Reroute If There's a Lot of Traffic Ahead?

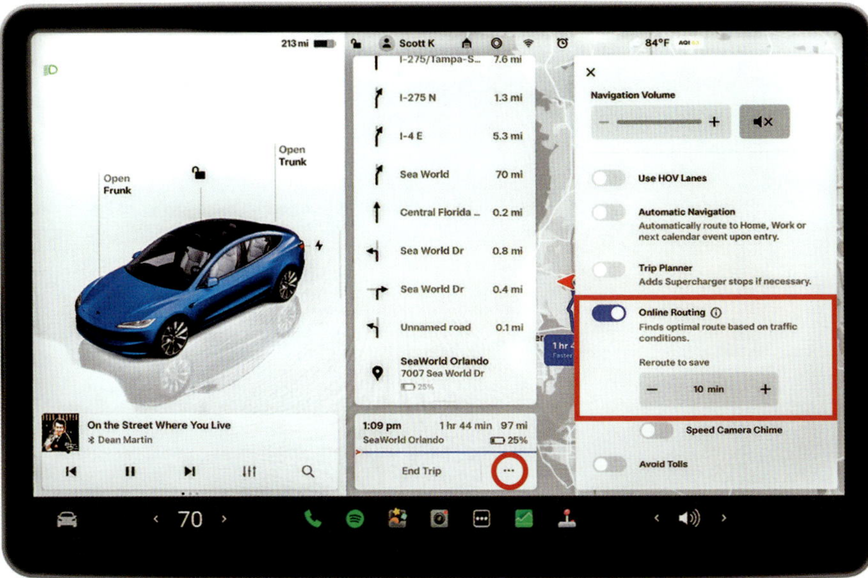

If your navigation system sees traffic backed up ahead that's going to impact your arrival time, there's an option to have it automatically reroute you around the traffic, and what's nice is, you get to decide how many minutes of delay it should be before this option kicks in. In other words, you're telling it: "If the delay looks like it's going to be 10 minutes or longer, then reroute me. Otherwise, keep the same route." To do this while you're navigating, tap on the three dots at the bottom of your directions list and in the menu that pops down, tap on the Settings icon (it looks like a gear). In the Settings window, turn on Online Routing (as seen here), and then below that, under Reroute to Save, choose how many minutes of delay you want it to be before it reroutes you around the traffic delay—you do this using the + and – buttons, and it displays the number of minutes in the center between those two buttons. You can also get to this Settings window by tapping on the Controls icon (the car icon in the bottom left of your center touchscreen), then tapping Navigation.

How Do I... Stop Having It Automatically Navigate Me to Work or Home?

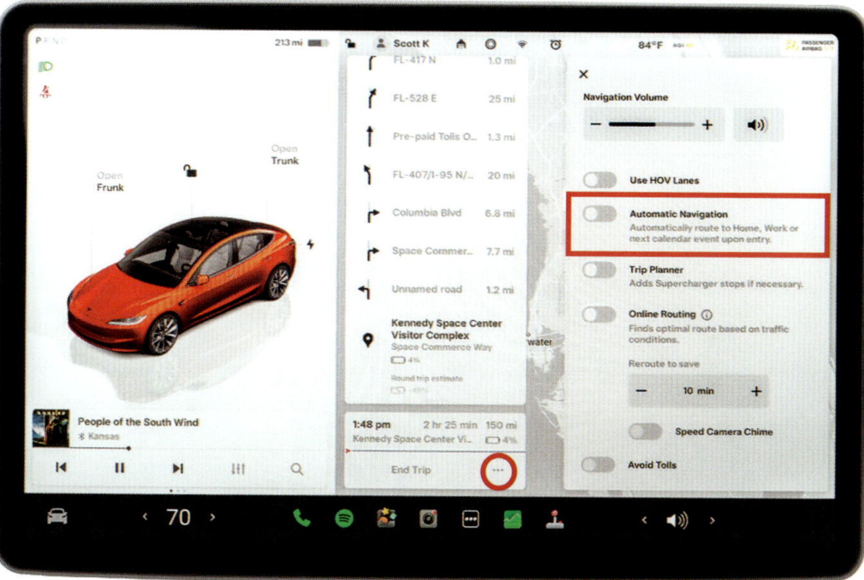

There's a feature that's either (a) really helpful, or (b) really frustrating, which is that your Tesla will assume that when you get in it in the morning you're going to work, so it automatically starts navigating you there. When you get in it at the end of a work day, it assumes you're going home, or if there's an event in your phone's calendar (and your calendar's synced to your car—see page 69), like "Dinner at The Cheesecake Factory at International Mall at 7:00 p.m.," it will automatically start navigating there if you get in anytime after 5:00 p.m. (a two-hour window. Of course, that would put me at The Cheesecake Factory at around 5:25 p.m., which is a dangerous place for me to be early). Like I said, this feature is either brilliant or maddening, especially if you've got the day off, or if after work you're heading to dinner and a movie, instead of home, so you wind up having to cancel routes all the time. You can turn this feature off, while you're navigating, by tapping on the three dots at the bottom of your directions list and in the menu that pops down, tapping on the Settings icon (it looks like a gear). In the Settings window, turn off Automatic Navigation (as seen here). You can also get to this Settings window by tapping on the Controls icon (the car icon in the bottom left of your center touchscreen), then tapping Navigation.

How Do I... Have an Address I Put into My Phone Automatically Go to My Nav System?

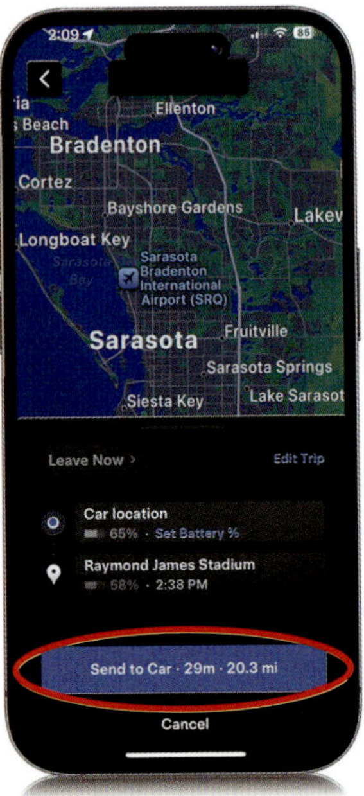

Open the Tesla app on your phone, tap on Location, and then in the Navigate search field, enter the address, or business, or point of interest you want to send to your Tesla. In this example, I entered "Raymond James Stadium" and it came up with its location. It shows how long it will take to get there, how many miles it is from where I am right now, and what the time and my battery percentage will be when I arrive. Now, just tap the blue Send to Car button at the bottom of the screen and when you get in your car, that address will already be entered into the navigation system for you.

How Do I... Know If I Need to Supercharge on My Road Trip?

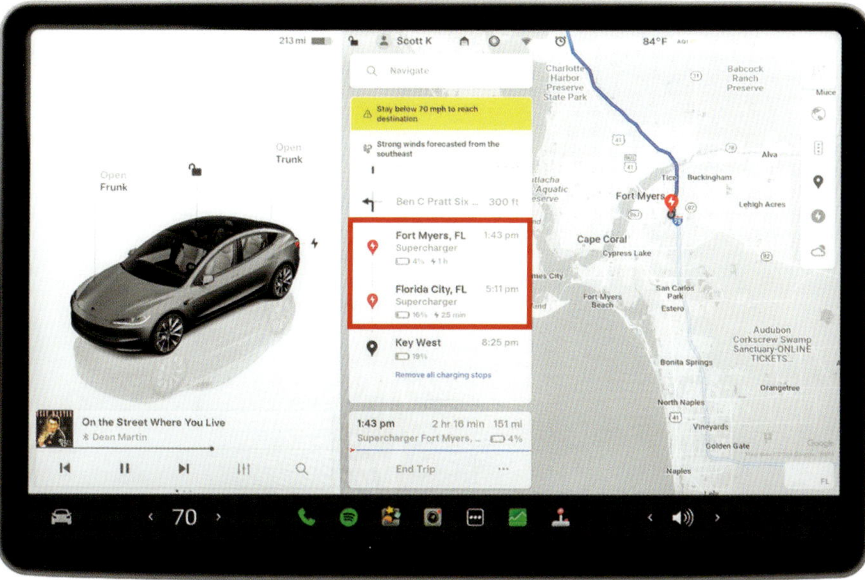

When you're taking a road trip, and you enter the address of your destination into your navigation system, if you won't have enough battery charge to reach it, the route displayed will have you stopping at a Supercharger (or multiple Superchargers, if it's a really long trip) along the way, and it will let you know how long you need to charge at each stop. For example, if I wanted to drive from my home in Tampa, Florida, down to Key West, it would automatically include in the route a couple of Supercharger stops along the way. The first would be in Fort Myers to Supercharge for one hour (and, luckily, this Supercharger is at a Wawa, so I can get one of their yummy sandwiches and perhaps a chip or two). The second Supercharger stop would be in Florida City, but for only 25 minutes.

How Do I... Know If I'll Have Enough Battery Left to Get Back Home?

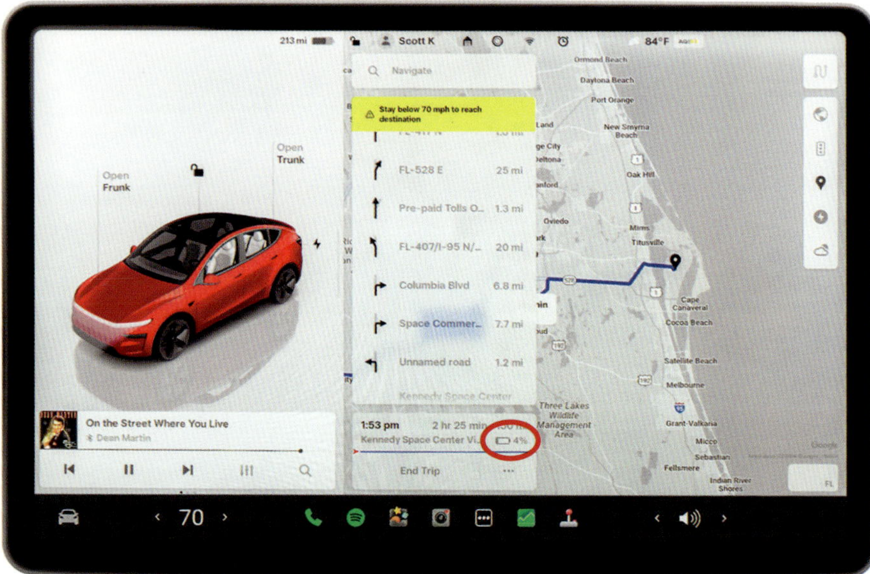

When you type an address into the navigation system and the list of your turn-by-turn directions pops down, at the bottom of that list, it shows you an estimate of how much battery you'll have when you arrive at your destination. But, more importantly, right below that it displays the round trip estimate of how much you'll have left when you make the trip back to where you started.

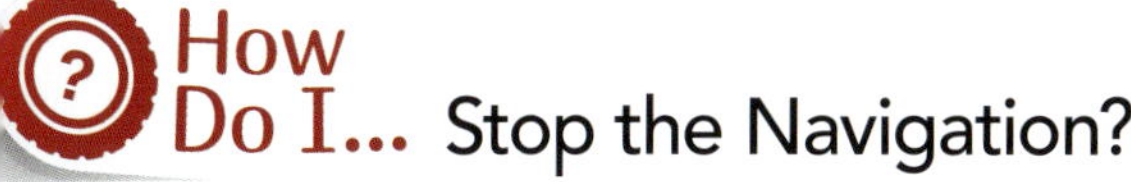

How Do I... Stop the Navigation?

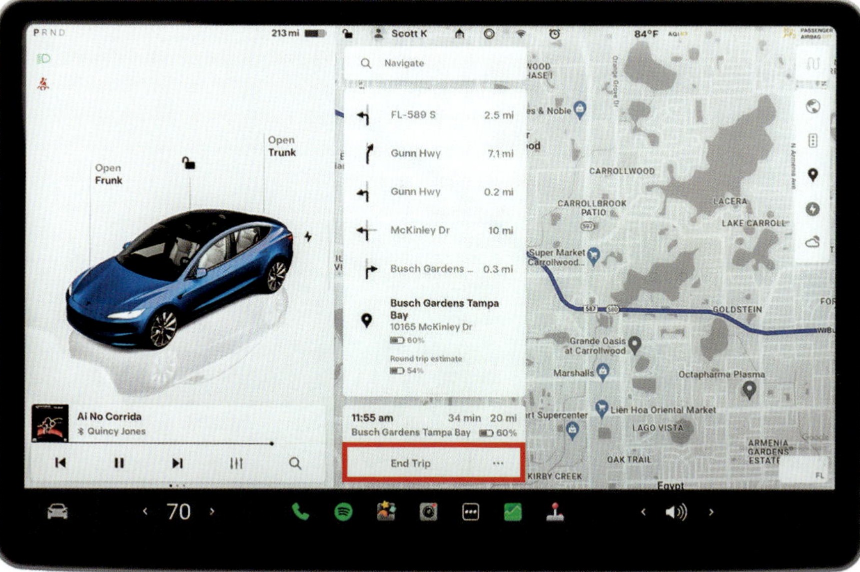

Under the Navigate search bar, you'll see a list of turn-by-turn directions for your current trip. At the bottom of the list, tap the End Trip button and it stops the navigation. You can also use a voice command—just press the Microphone button (or right scroll button in older models) on the right side of your steering wheel, then release it and say, "Stop navigation." That's all it takes.

How Do I... Call a Business I See on My Map?

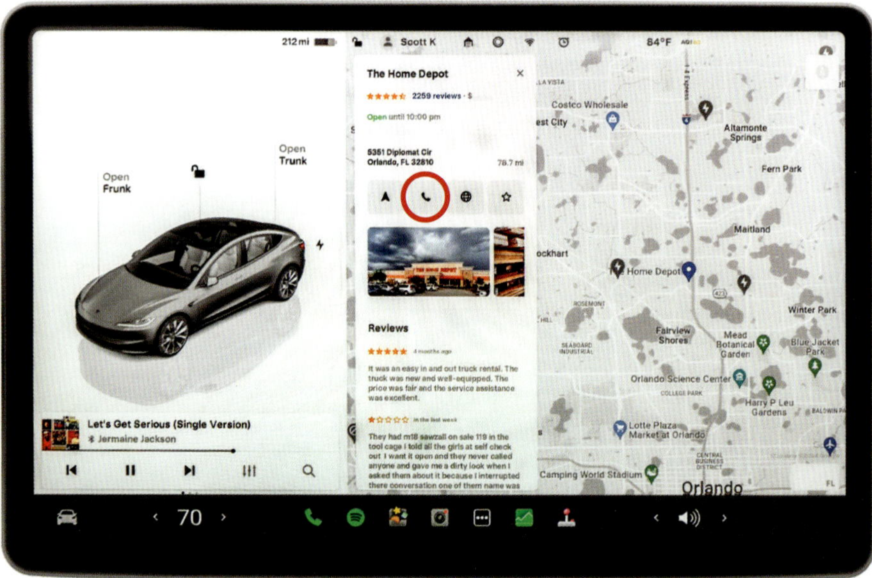

This is a very "Google Maps" type of feature. If you locate a business by searching directly on the map—for example, let's say you found The Home Depot—when you tap its pin, a card pops up with things like their address, rating (1 to 5 stars), hours of operation, a button you can tap to visit their website, and a button you can tap to call them, all right from the map. How awesome is that?

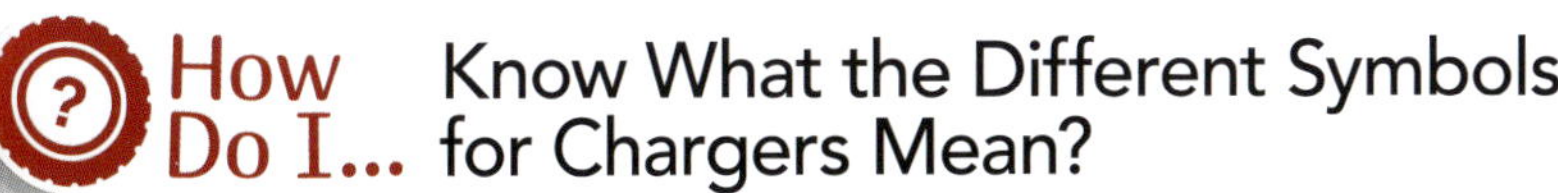

How Do I... Know What the Different Symbols for Chargers Mean?

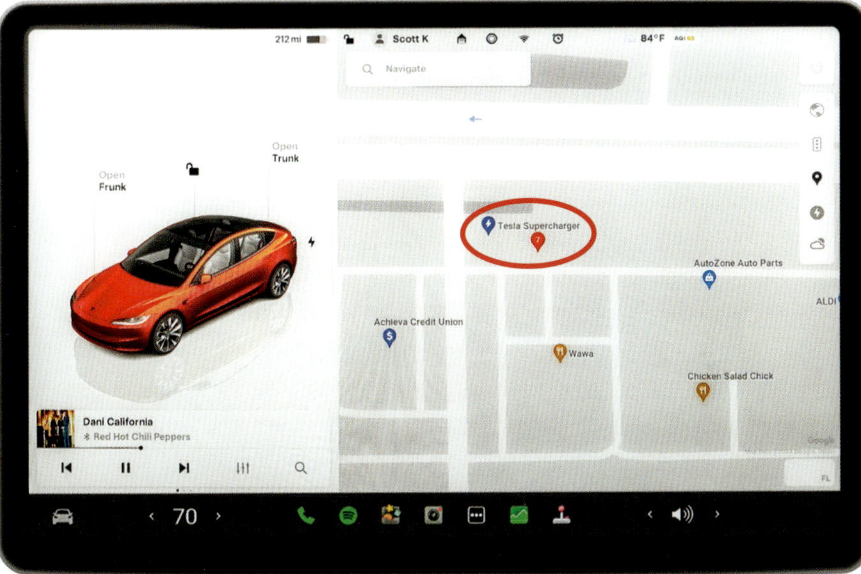

If you tap on the Navigate search bar, a menu pops down below it, and one of the tabs here is called "Chargers." If you tap on this, it displays pins with little lightning bolts on the map showing places you can charge. Some of them have pins with different symbols on them because they are getting live updates from the Superchargers themselves (which is pretty amazing, if you think about it). If you're wondering what those different symbols mean, well here ya go: A pin on the map with a "no" symbol on it, means that Supercharger is closed (rare, but it happens). A wrench icon lets you know the Supercharger is working at a "reduced capacity." A clock icon lets you know that the Supercharger is busy, so you might have to wait for a stall to open. If a pin is red with a lightning bolt icon, that means that particular Supercharger isn't communicating with the map at this moment, but it most likely is okay. If a red pin has a number on it, that's the number of charging stalls that are currently available (as seen here, where this location has seven stalls available). If the pin you see is black, that lets you know it's not a Supercharger, but more likely is a Destination Charging location (one that charges at a similar speed to a wall charger you'd have at home).

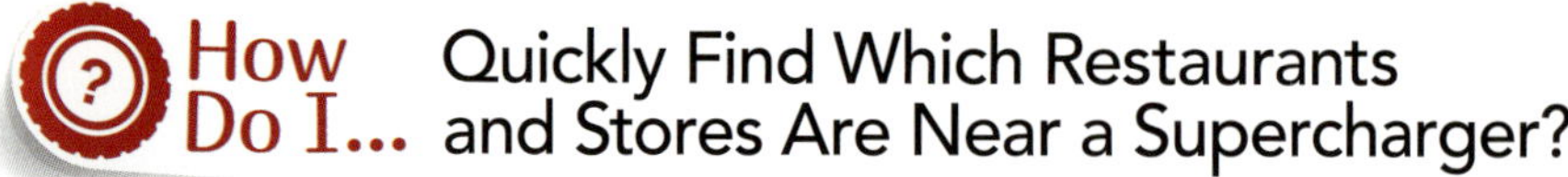

How Do I... Quickly Find Which Restaurants and Stores Are Near a Supercharger?

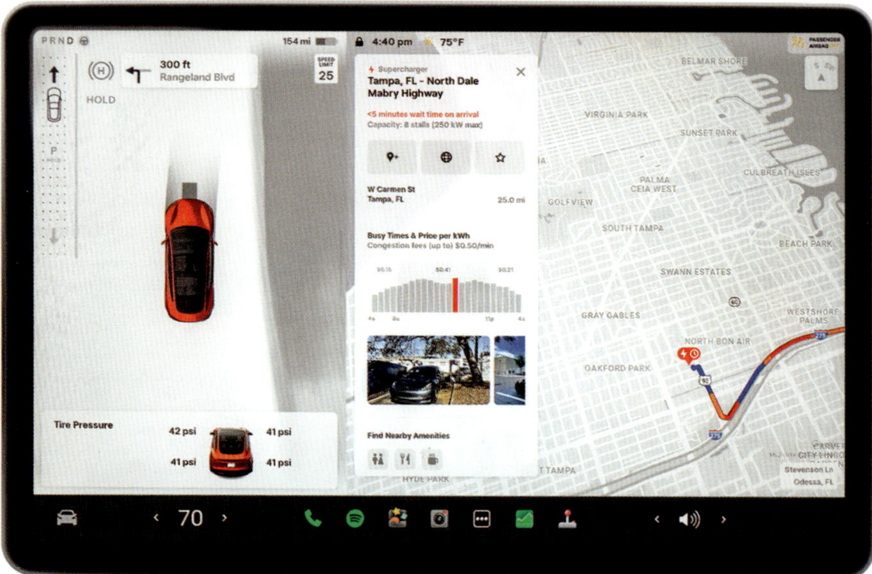

When you find a Supercharger on the map and you tap on its pin, a little card pops up with lots of information about that Supercharger (like how many stalls it has, the current wait time, etc.), but you'll notice there are also little icons for food, restrooms, and shopping at the bottom. Those aren't just icons; they're buttons. If you tap on any one of those, it will bring up a list onscreen of which restaurants (or stores) are nearby, so you can make plans for what you want to do while you're charging before you even arrive. (*Note:* See Chapter 9 for more on using Superchargers.)

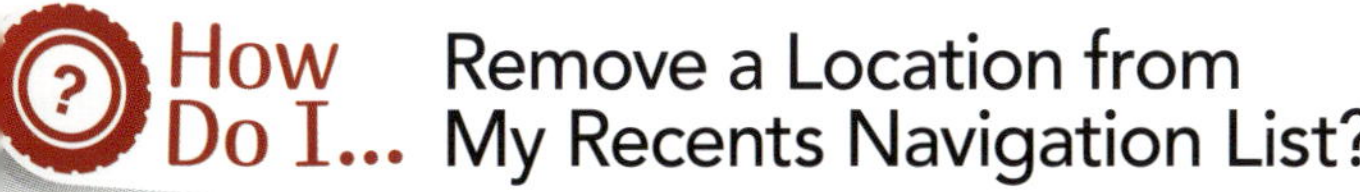

How Do I... Remove a Location from My Recents Navigation List?

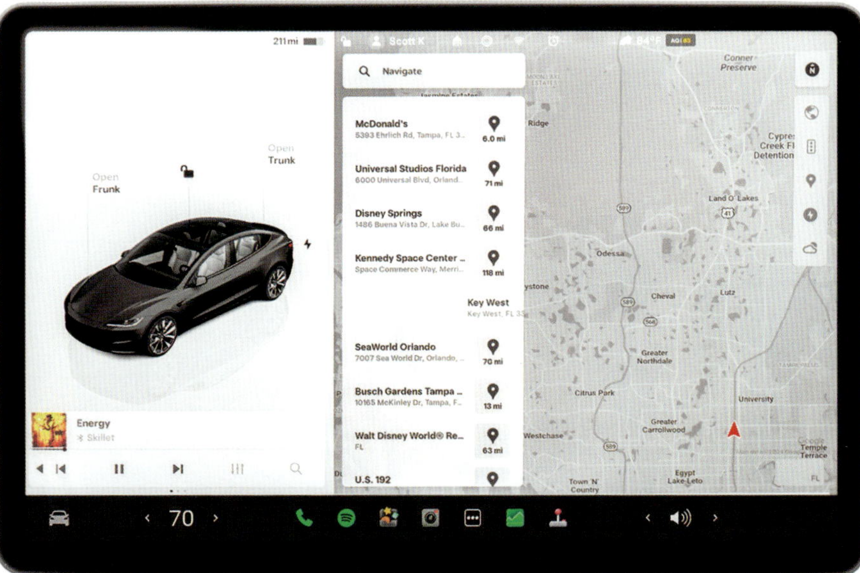

Let's say you went to a particular store to buy a surprise gift for your spouse and you don't want that location to appear in your recent locations navigation list, so you don't tip them off. You can erase any location from the list by tapping on the Navigate search bar, tapping Recent, and then simply swiping right on the location in the list and it's gone. (*Note:* I should get five bonus points for coming up with a reasonable scenario where you'd want to wipe a location from your history that didn't have a nefarious nature. Just sayin'.)

Chapter 7

How to Use the Tesla App

Yes, There's an App for That

If you haven't downloaded the free Tesla app yet, you should do it right away because it's the springboard to a cornucopia of wonders that dare not speak its name. The app is very mature at this point (so it uses a lot of unsavory language), and it lets you control so many things in your Tesla—everything from opening the trunk or the frunk, to starting the air conditioning (or heating) before you get in, and a host of other really helpful things (I know this is bordering on something I said I wouldn't do—which was to share useful information in the chapter intro. But, don't worry, it gets back to being meaningless in just a moment). Anyway, when the app first came out in the App Store, it was not a hit at all. In fact, the overall rating for a long while was just 1 star, which I didn't think it deserved, so I went and looked at some of the reviews. The first one I read said, "I don't have a Tesla, so this app is stupid." Well, I can't argue with that because the app is only for Tesla owners, and if you don't have a Tesla, well…it probably is stupid. I felt, as a responsible Tesla owner and educator, I should drop in there and give some context to the review, so I posted this: "Yeah, and your face is stupid, you big stupid head." Well, let me tell you, that shut him right up. Not a peep back because that was such a super-burn that there's really no coming back from it. Not only did he not respond, he deleted his review, and his account, and as best as I can tell, he also sold his computer as well. #roasted! That's right; he never posted a reply again. However, one day, when I needed to get my gutters cleaned, I went to that app TaskRabbit—where you hire people to do repairs and stuff around your house—and I hired a guy whose bid was way lower than any others I got on the app. Anyway, the guy shows up at my house, climbs up the ladder, and he's all like, "Oh, you're fine with me cleaning your gutters, but you think my face is stupid." It was at that moment I realized, holy cow, this was the guy who wrote the Tesla app review. He gets this really smug look and says, "Well I think your face is stupid," and you can tell he's about to go off on me, but then he looks down at me and says, "Ummm, is that a TASER 2 Pulse?" and, spoiler alert, I never heard from him again. The end.

How Do I... Open My Trunk or Frunk?

One of the most common things Tesla drivers want to do from the Tesla app is open and close the trunk and open the frunk (front trunk). If you have a late model Tesla, you can open and close the trunk from your app. But, keep in mind that if you open the frunk on any Tesla (besides the Cybertruck), you have to close it physically by pressing it down. There's no way to close it from the app. In the Tesla app, tap on Controls and you'll see a top-down view of your car. You can then tap on Open that appears on the frunk and the trunk (as seen here). When you open the trunk on a Tesla with a power close (again, 2021 models and newer), you'll then see Close over the trunk, which you can tap to remotely close it. And, once again, if you tap Open on the frunk, you'll have to physically close it by hand (unless you have a Cybertruck).

How Do I... Lock/Unlock My Doors?

In the Tesla app, tap on Controls and you'll see a top-down view of your car. Right on the top of it, you'll see a lock icon (circled here), and it will either appear as locked or unlocked, showing the current lock state of your Tesla. Just tap the lock icon to lock or unlock it. One thing to keep in mind is that using the Tesla app to control your Tesla remotely is heavily dependent on your Tesla being able to connect to the internet wirelessly. This means that if it's several levels deep in a parking structure, you may not be able to control it via the app.

How Do I... Start My Tesla?

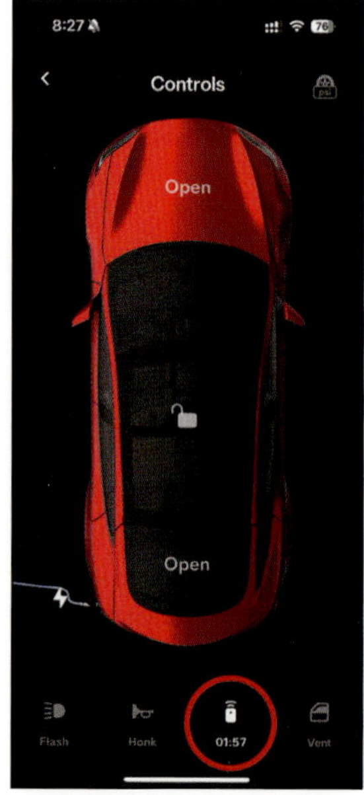

Sometimes you may need to start your Tesla remotely to allow someone to drive it that doesn't have your key card—perhaps it's a family member that needs to move your car while you're not home. From the Tesla app, you can unlock the doors so that they can get in, but you can also start it, so that they can drive it. Tap on Controls, and then tap on the Lock icon if you need to unlock the doors. Once they're in your car, tap the Start icon, and they'll then be able to drive. They will have two minutes to get in and start driving before the remote start cancels (you'll see a timer onscreen counting down). Once they put the car in park and exit, you would need to remote start it again for them to drive it again.

How Do I... Manage My Tesla Keys?

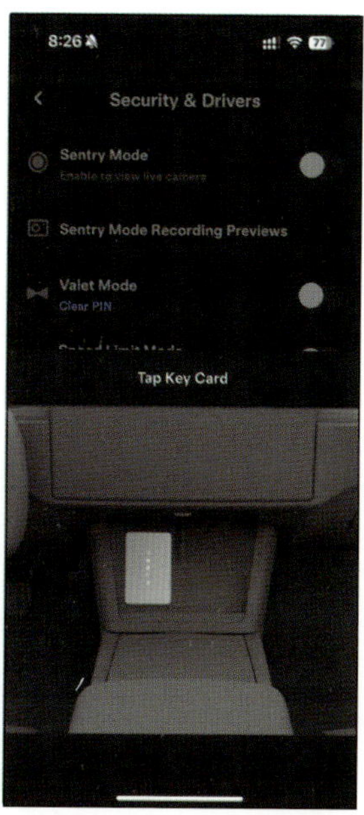

Late model Teslas all use a smartphone (Android and iPhone) as a phone key. This means that as you approach your Tesla, it will automatically unlock, even if your phone is tucked away in your pocket/purse. You just get in and your Tesla is ready to drive. You have the option to turn off Phone Key, but I can't think of a reason why you would. You also got a physical key card (or key fob on older Teslas) when you got your Tesla. These physical key cards can unlock your Tesla by positioning them on the B-pillar beneath the camera between your front and back driver's side doors. Once you get in your Tesla, place your key card on the center console near the arm rest (or on the phone charger, depending on your model) to activate your Tesla for driving (see page 12 for more on this). You can add new key cards in the Tesla app (see page 2 for more on this), but you will need to get inside your Tesla with a working phone key or key card to add the new one to your Tesla. It will ask for an existing key card in order to activate the new one. If you lose your key card, you can deactivate it in the Locks section of your center touchscreen and you can order additional key cards from the Tesla online store.

TIP: CARRY A BACKUP KEY CARD

Always carry a working key card with you as a backup in case something happens to your smartphone.

How Do I... Warm Up/Cool Down My Tesla Before I Get In?

Since your Tesla is an electric vehicle, there is no engine. This means that you can turn on the climate controls without having to start it. You can control the climate of your Tesla to warm it up or cool it off before you get in by opening the Tesla app, tapping on Climate, setting it to your desired temperature, and then tapping the On icon. You can also defrost your car this way in cold climates. Although you can easily turn on your climate remotely to warm up or cool off your Tesla before you get in, there is an energy impact on your battery charge. For most people, most of the time, this won't be a big deal because you're just turning it on for a few minutes before you're going to get in anyway. However, if your state of charge is 20% or less, then you may not be able to turn on the climate remotely. Tesla automatically disables these features to make sure that you have enough of a charge to make it to your destination or a charger. There is a 20% state of charge safety threshold.

How Do I... Not Forget to Turn on Dog Mode?

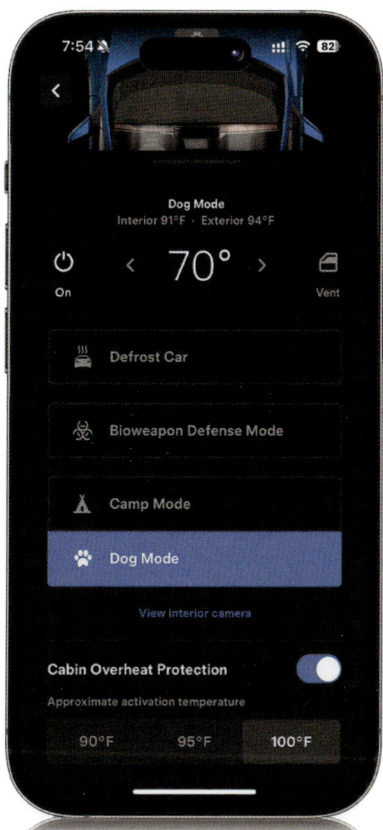

Your Tesla has multiple modes to leave the climate settings turned on whether you're in your car or your precious loved ones are—i.e., your pets. For example, there's a Camp Mode for when you want to sleep in your Tesla and have the air conditioning or heat on. You simply enable this mode from the Climate controls at the bottom of your Tesla's center touchscreen (see page 57 for more on Camp Mode). However, if you left your pets in the car and forgot to turn on Dog Mode, thankfully, you can do it right from the Tesla app without having to run back to your car. Just tap on Climate, and then scroll down and tap on Dog Mode to turn it on. The mode's button will turn blue, indicating that it has been enabled (as seen here). You can also set the temperature at the top of the screen and you'll see the air conditioner animation up there as well. The Climate icon will switch from Off to On and there will be a large message on your Tesla's touchscreen, letting people that pass by know that there is no need for concern because your pets are nice and cool (or warm in the winter). You can even view the interior camera to check in on your pets from time to time.

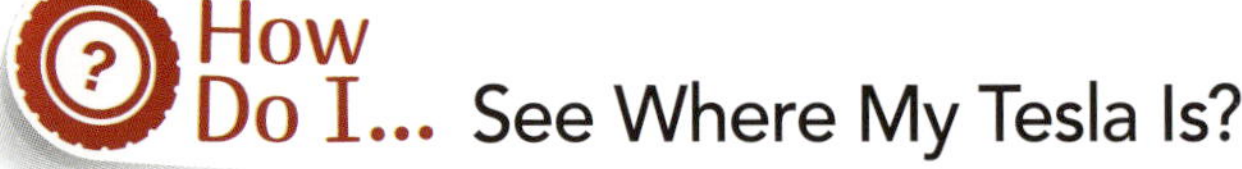

How Do I... See Where My Tesla Is?

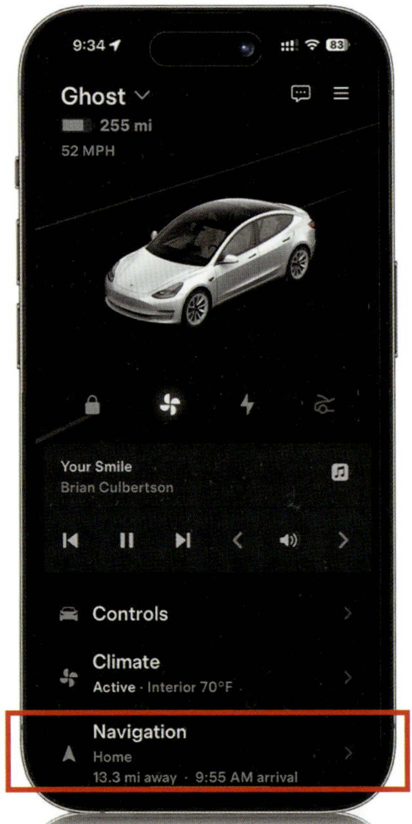

If you let a friend or family member drive your Tesla and you want to see where it is, just tap on Location in the Tesla app, and then tap Navigation, and it will show you exactly where your Tesla is. If it's moving, it will show you details from the navigation, like how far away from home it is and its ETA. If you want to mess with the driver, you can skip songs that are playing or turn up the heat. Just kidding, but you could.

How Do I... Remember to Plug in My Tesla?

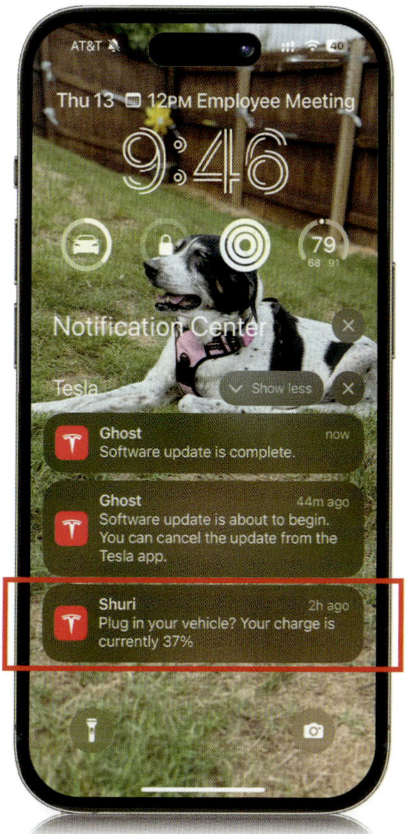

Every Tesla owner has probably forgotten to plug in their Tesla before going to bed at one time or another. This may be no problem at all if you have plenty of charge and a short commute the next day. However, if you are in a hurry and don't have enough of a charge to get through your day, you may end up having to stop at a Supercharger for a few minutes (see page 177). Luckily, if you have a low state of charge, the Tesla app will send you a reminder to plug in.

How Do I... Plan My Road Trip?

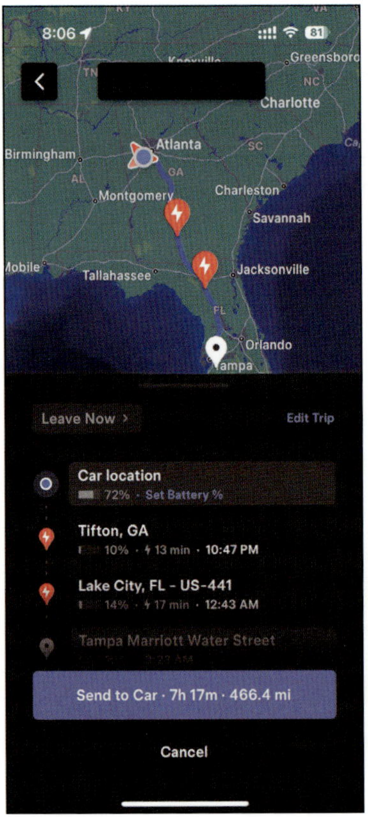

One of the things you're always going to want to know is if you can make it to your destination without charging or, if you can't, where will you need to stop and charge. This is called EV-trip planning, and luckily, it's built into Tesla's app as well as your Tesla's center touchscreen (see page 159). Tap on Location in the app, then type in the address or name of the location you're driving to, and you'll get the route and any necessary charging stops based on your state of charge. You can change the starting charge level to get a more accurate estimate if you plan to charge before you leave (see page 107). I usually charge to 100% for road trips before heading out.

How Do I... See Available Superchargers?

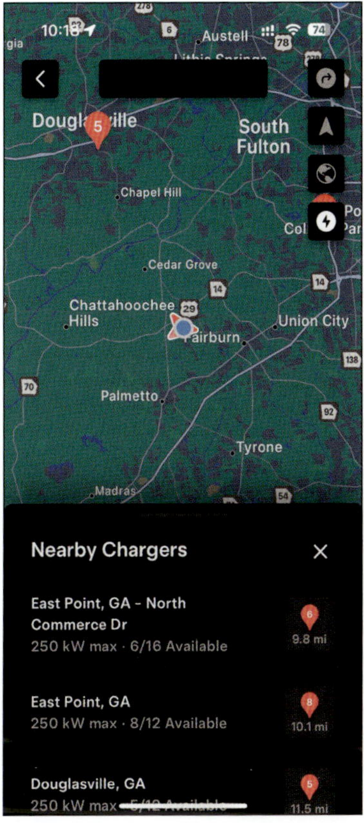

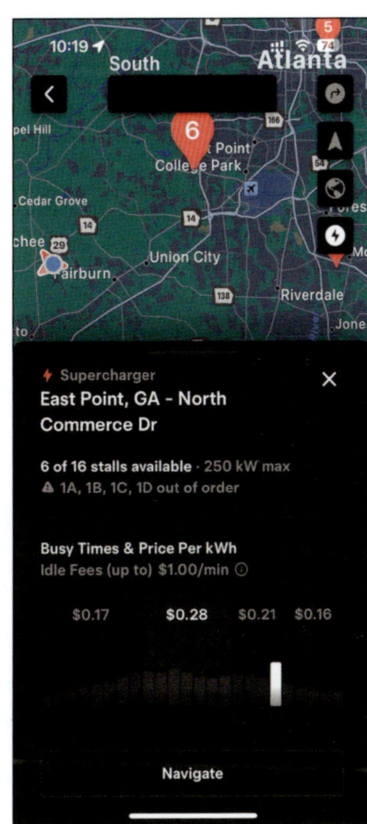

Even if you charge at home or at work, there may come a time when you need to find a Supercharger. To see nearby Superchargers in the Tesla app, tap on Location and then tap on Chargers. A list of the closest Superchargers will come up and not only will you see where they are, but you'll see how many stalls they have and how busy they are. If you tap on one of the Supercharger locations, you'll see more details, like the speed of the chargers, as well as the kWh cost. You can then tap the Navigate button to get to the one you want and it will send this information to your Tesla.

How Do I... Set Up and View My Charge Stats?

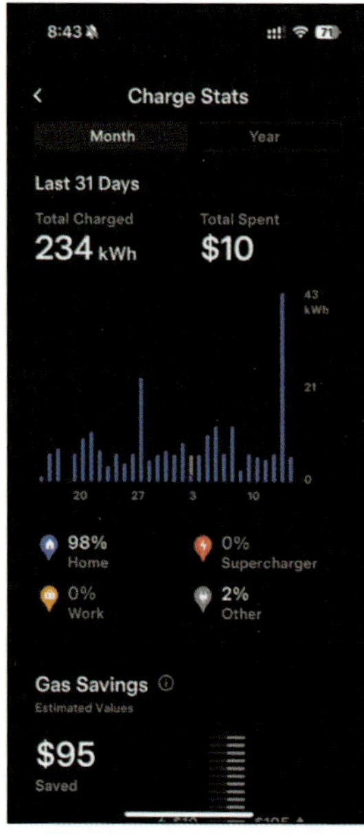

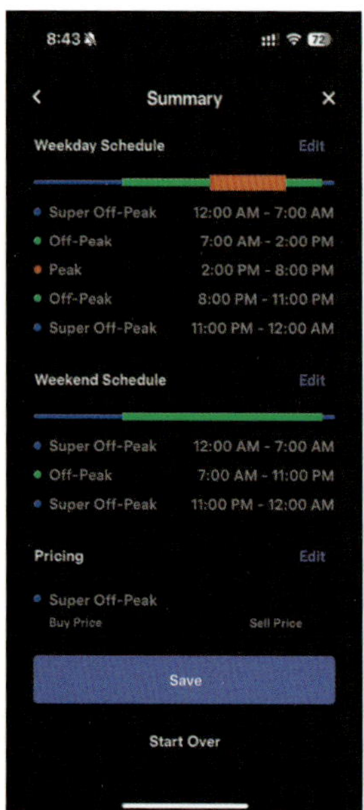

One question you'll definitely get from friends and family is, "How much did your electric bill go up?" The Tesla app has Charge Stats built in, and you can even put in your utility electricity rates, so that you get a more accurate estimate of what your Tesla costs to charge. Tap on Charge Stats, then tap on Settings, and you'll see three sections: Home, Work, and Other. Tap on Home and you can put in your utility provider and the rates you pay. (Since your Tesla already knows how much your Supercharging was, it will automatically put those costs in.) Once you set up your rates, you'll then see approximately how much it costs to charge your Tesla each month and how much you potentially saved vs. buying gas. You can check your stats anytime after you charge, but keep in mind that it's just an estimate and doesn't take into account things like solar energy discounts if you have solar panels or a solar roof. The estimate assumes that you charged your Tesla via your home's electricity rates. While we're on this subject, you should check with your utility company to see if there is a TOU (Time of Use) rate for your home. This allows you to pay the lowest rates at night (see page 111 for more on this). You can schedule your Tesla to only charge during these times, even though you plug it in when you get home.

How Do I... Set My Tesla's Security Features?

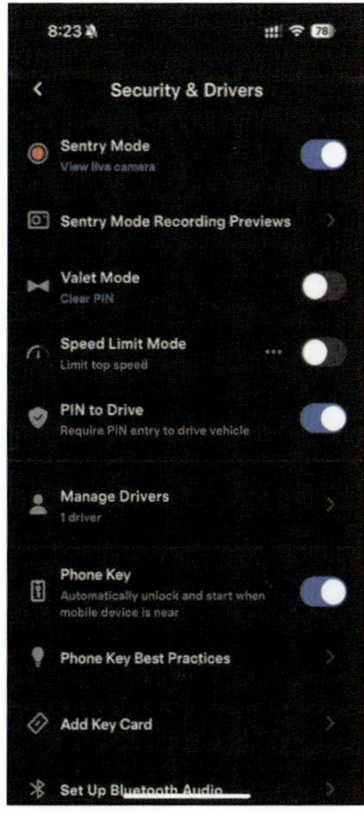

There are several options for security that you can access right from the Tesla app by tapping on Security & Drivers. Your first option is Sentry Mode and turning this on allows your Tesla to automatically record suspicious activity happening around it. For example, if someone comes close to your parked Tesla, your cameras will record them. If they start trying to open doors or do things to it, a loud alarm will sound and the touchscreen and lights will start flashing. Sentry Mode does have an impact on your charge, so only turn it on as needed. If you valet park, you should enable Valet Mode and set a PIN to disable it. You will need to give the valet your key card/key fob to be able to drive. Turning on Valet Mode will hide your Navigation home/work favorites from your touchscreen and it will limit how fast your Tesla can be driven to 70 mph. It will prevent access to the frunk and glove box, and voice commands are disabled, as well as Autopilot features and HomeLink (if your Tesla has this option), so they can't open your garage. Some entertainment apps are also disabled—settings that a valet should not be changing or accessing. If you forget the PIN to disable Valet Mode, you can reset it in the app using your Tesla username and password. Speed Limit Mode is just as the name implies. It allows you to set a maximum speed limit that your Tesla can be driven—this is great for when you let your teenagers drive. Another great security feature that I highly recommend enabling is PIN to Drive. With this enabled, a 4-digit PIN needs to be entered on your center touchscreen before your Tesla can be driven. This means that even if someone has your key card or phone, they still can't drive without the PIN.

How Do I... View My Sentry Mode Cameras?

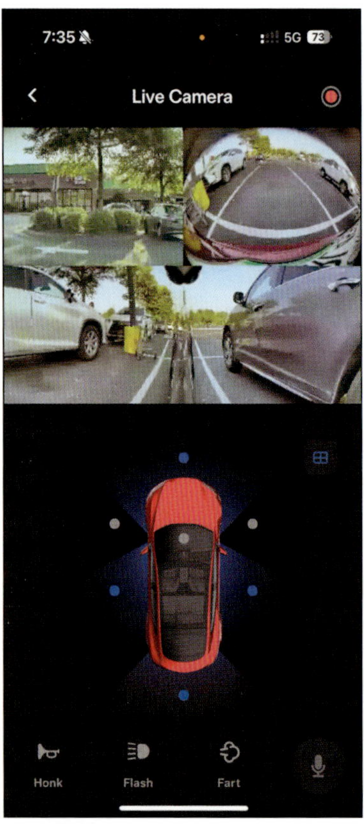

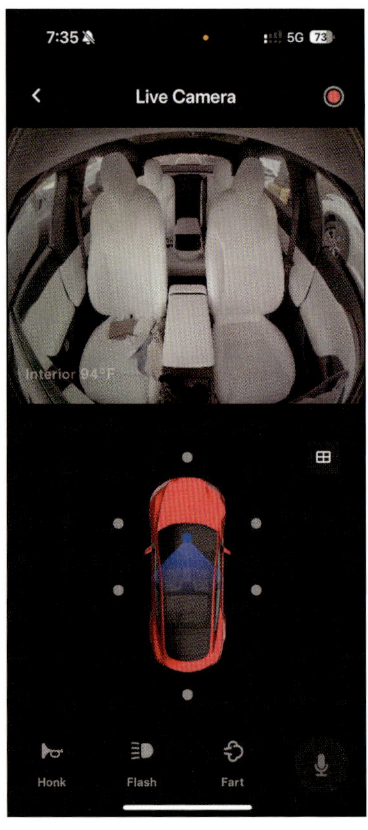

Your Tesla uses its external cameras for Sentry Mode and dashcam recordings, and you can see these cameras, along with its interior camera (if your Tesla has one), any time your car is parked and no one is in it. Sentry Mode will need to be enabled in order to see the cameras in the Tesla app, and thankfully, you can enable it from the app's Security & Drivers settings. Once Sentry Mode has been enabled, you can tap Live View & Cameras to see your Tesla's surroundings. By default, it will show all but two of your external cameras. For some reason, the side B pillar cameras (in between the front and back windows) are a separate view, and you can tell that they are not being viewed as they will be gray icons on the left and right side of the picture of your Tesla. But, you can tap either side to see those cameras (as seen here, on the left, where their dots are blue). When you want to see the inside, just tap on the inside of it (as seen here, on the right).

How Do I... See Sentry Mode Notifications?

Sentry Mode is Tesla's built-in alarm system that records the things that are going on near your Tesla when you're away from it. Normally, it just starts recording any motion events that take place near your Tesla and doesn't bother alerting you of those type of events. However, if someone tampers with your Tesla and triggers the actual alarm, it will not only record and sound the alarm, but it will also notify you via the Tesla app. When that happens, you'll be able to view those Sentry Previews by tapping on Security & Drivers, and then tapping on Sentry Mode Recording Previews. If there were any serious Sentry Mode events, they would be here for you to review.

How Do I... Talk to Someone Standing Near My Tesla Remotely?

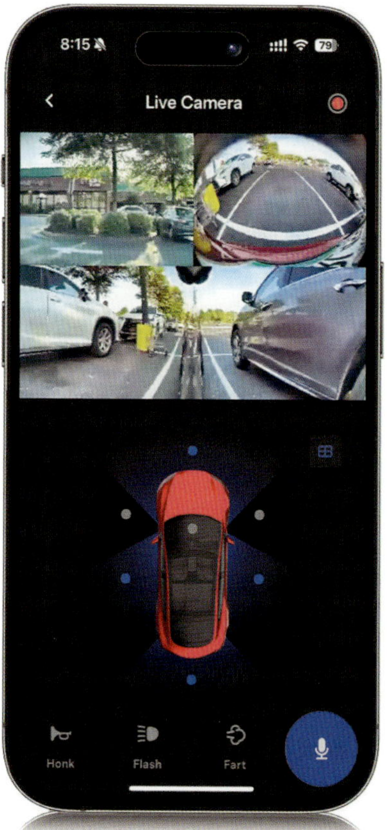

If your Tesla is equipped with the pedestrian speaker, you can actually use that speaker as a megaphone and use your Tesla app as the microphone. Go to Security & Drivers in the app and make sure Sentry Mode is enabled, then tap Live View & Cameras to see camera view. Next, tap the Microphone icon to make an announcement through the pedestrian speaker. It's a fun time, especially if the people around your car aren't expecting it.

How Do I... Flash My Lights or Honk My Horn?

We've all been there. You go out to a parking lot or parking structure and you can't quite remember where you parked your car. Luckily, you can use the Tesla app to either flash your lights or sound your horn to help you locate it. The next time you need to find your Tesla, open up the app, tap Controls, and at the bottom of the screen, you'll see a Flash icon and a Honk icon. Just tap either icon for whichever one you want to do.

How Do I... Check My Tire Pressure?

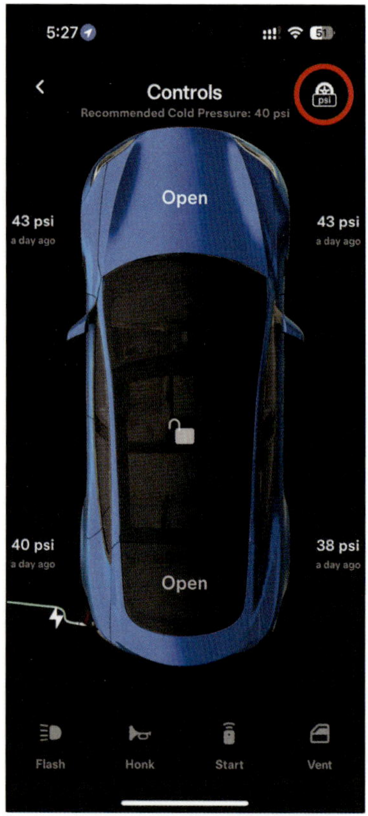

One of the great things about having a connected car like Tesla is that you don't have to go out to the car to get stats, like tire pressure. You can check your tire pressure any time right from the Tesla app on your smartphone. Launch the app, then tap on Controls, and then tap on the Tire Pressure icon in the upper-right corner to see the current pressure of each tire. Looks like I'd better put some air in my passenger side rear tire.

How Do I... Check My Odometer, VIN, or Software Version?

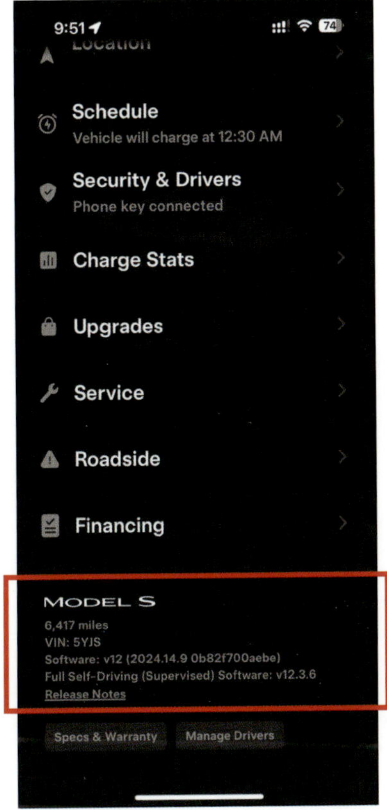

Usually, in order to see how many miles your car has on it or to see the VIN (vehicle identification number), you would have to look inside your car. Luckily, this information can be viewed right in the Tesla app—it's all right there at the bottom of the main screen. You'll also see the current software version that is installed in your Tesla, as well as a link to read the Release Notes. (I'm so glad that they finally added this link to the Release Notes because, back in the day, you'd have to go read them on your Tesla's center touchscreen after the update was installed.)

Schedule a Service Appointment?

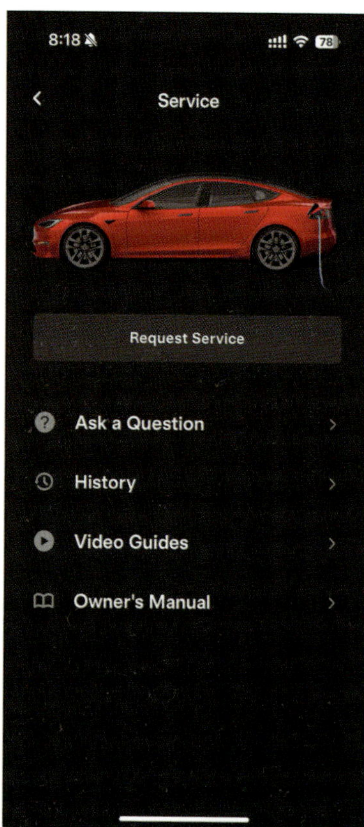

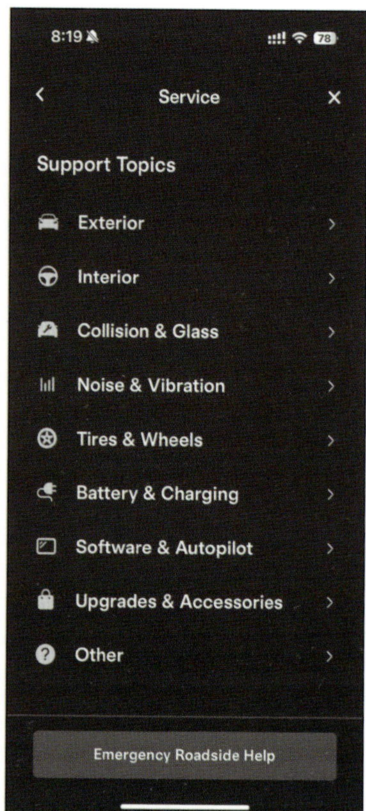

This can be done right in the Tesla app by tapping on Service, and then tapping on the Request Service button. You'll then pick a category for the type of service you need, and from there, follow the prompts to describe the issue and even upload photos if necessary. Depending on the type of service you need, Tesla may be able to dispatch a mobile tech to come to you. This is extremely convenient because you can receive service from the comfort of your driveway or work parking lot. The Tesla service center will reach out to you via Messages in the app to confirm your appointment, provide an estimate, or collect any other information they may need.

How Do I... Request Roadside Assistance?

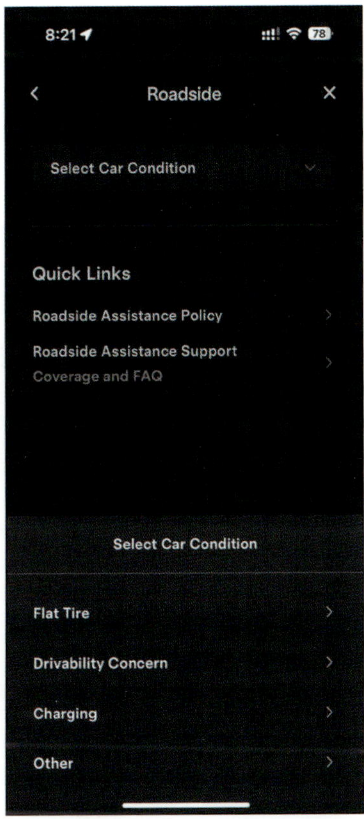

If for any reason your Tesla is not drivable and you need help or even a tow, you can request emergency roadside assistance right in the Tesla app. Tap on Roadside and you'll then be prompted to select your car's condition—your choices are Flat Tire, Drivability Concern, Charging, or Other. You'll confirm other details of your problem, the location of your Tesla, and from there, a roadside assistance agent will reach out to you to discuss your options. If they can solve the problem remotely, they will. If they can send a tech out to do a repair, they will. Lastly, if need be, they will set up a tow truck to pick up your Tesla and tow it to the nearest Tesla service center. (*Note:* See more about service in Chapter 11.)

How Do I... Reset My Driver's Profile?

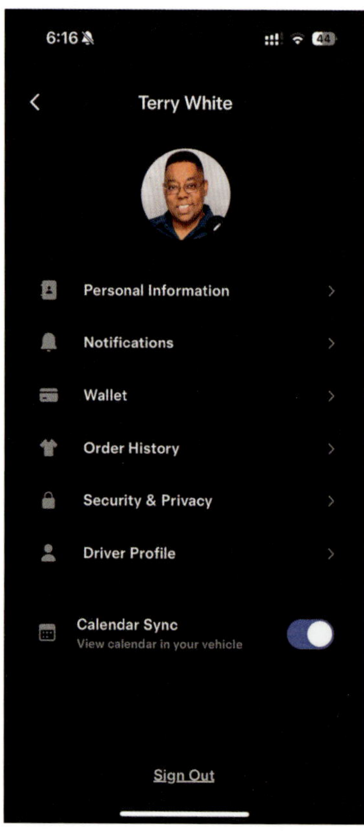

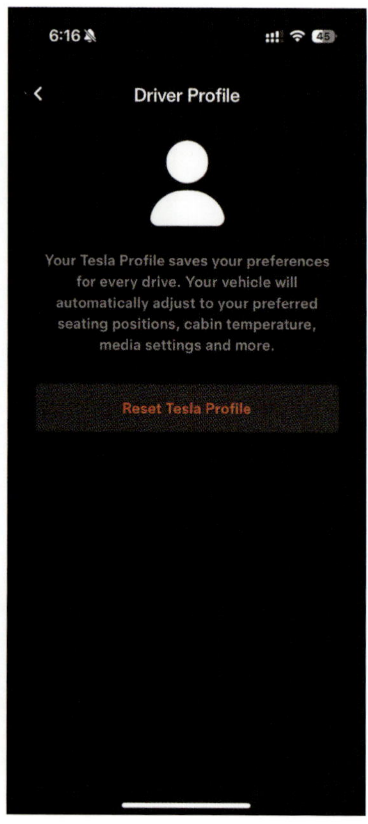

One of my favorite Tesla features is the driver's profile that syncs to/from your Tesla. When you adjust your seat position, cabin temperature, media settings (such as Spotify), and more, those settings sync to the cloud. This means that if you get in a different Tesla and sign in, your saved settings will be there. This was extremely cool when I rented a Tesla on a business trip and not only did my seat adjust, but my favorite music started playing. It was truly awesome. However, if you want to start from scratch, you can reset your driver's profile from the Tesla app. Just launch the app, tap the three-line (hamburger) menu icon in the upper-right corner, and then tap your name to see your driver's profile. If you want to reset it, tap on Driver Profile, and then tap the Reset Tesla Profile button. (*Note:* See page 10 for more on driver profiles.)

How Do I… Access My Owner's Manual?

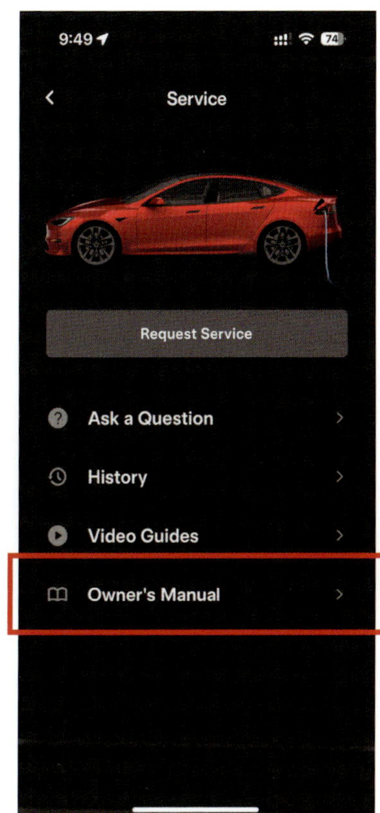

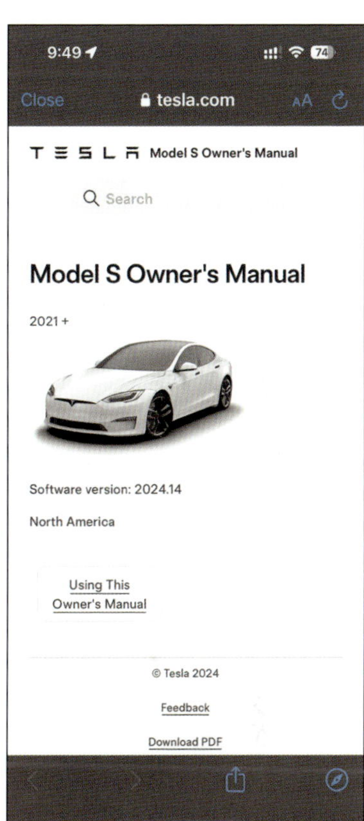

This book is awesome, right? However, things could change after the time this book was written. A software update could move things around, add new features, or change the way existing features work. Luckily, with each software update, Tesla updates the owner's manual and makes it available to look things up right in the Tesla app. Once you launch the app, tap on Service, then tap on Owner's Manual. A webpage will load right in the app with the most current manual for your Tesla. There's a Search bar right at the top, so that you can search for what you're looking for instead of having to scroll every page. When you're finished, just tap Close to close this webpage and get back to the app. (*Note:* We've also got you covered with new feature releases on the book's companion webpage. See page xix in the book's introduction for more on this.)

How Do I... Start a Pending Software Update?

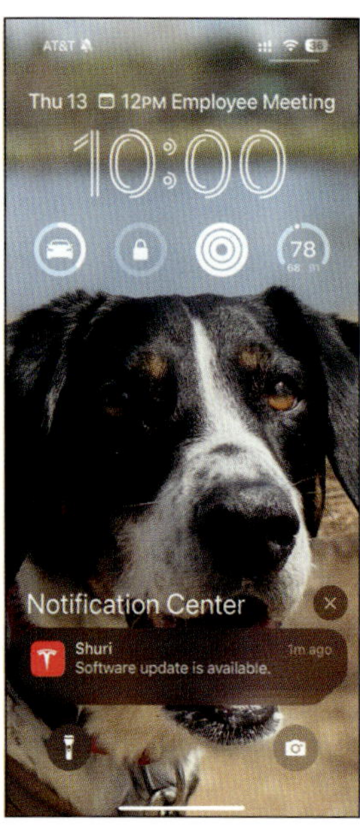

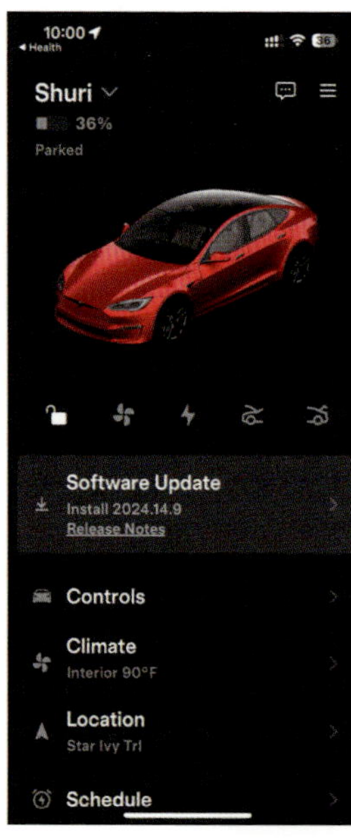

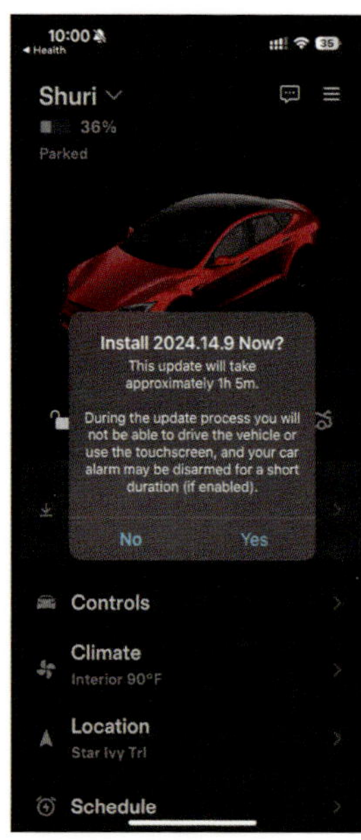

From time to time Tesla pushes out software updates for your Tesla, but it doesn't install them automatically. These updates can take anywhere from 20 minutes to an hour to perform and you would probably be upset if an update was in progress when you needed to go somewhere. This means that if there is an update that has downloaded to your Tesla and is ready to install, you have to actually initiate or schedule the update to take place at a time that you won't be using your Tesla (like at night while you're asleep). If there is a software update waiting, you will see it in the Tesla app and you can read the release notes first (see the next page). At a time that is convenient for you, tap on Software Update, and then tap on Yes to start the install. While your Tesla *does* need to be connected to WiFi to download the update, it does *not* need to be connected to WiFi to install it. People that don't have WiFi near where they park their Teslas sometimes use public WiFi or mobile hotspots to get the update, and then drive home to actually do the install. I also recommend only installing these updates when you're at home because although it's very, very, very rare, software updates could actually disable your Tesla due to some problem installing them. You wouldn't want to be stranded away from home if this happens because you cannot drive your Tesla until the update is finished installing. You also can't be Supercharging while an update installs. Your Tesla will send you notifications when the update is done installing.

How Do I... Read Software Update Release Notes?

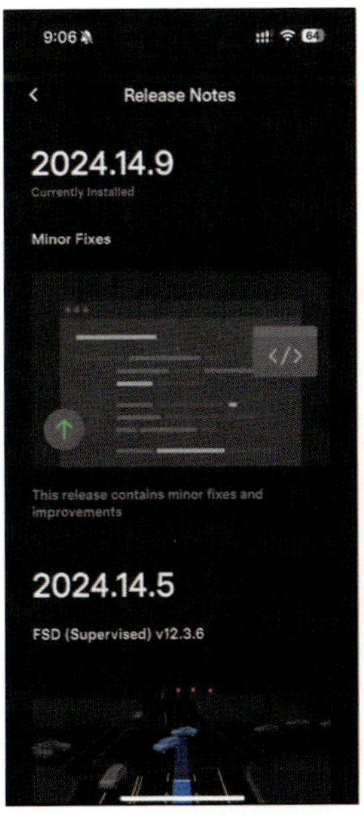

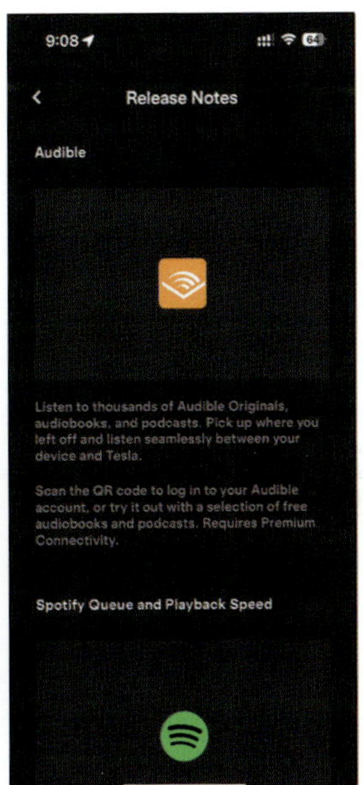

Your Tesla gets regular software updates that fix bugs and even give you new features. Most of the time your software updates download automatically if your Tesla has a WiFi connection. Although it downloads the update to your car, it doesn't install them automatically. This is because the updates can take several minutes to install and you should always be in control of when that update happens (see the previous page for more on this). Before you install an update, or to decide when you want to do it, you can check the Release Notes first to see which new features or fixes are in it. While you need a WiFi connection to download the update, you can install it at anytime without your Tesla being connected to WiFi. To read the Release Notes, open the Tesla app and scroll to Software Update, where you'll see a link you can tap to read them.

How Do I... Use My Control Buttons?

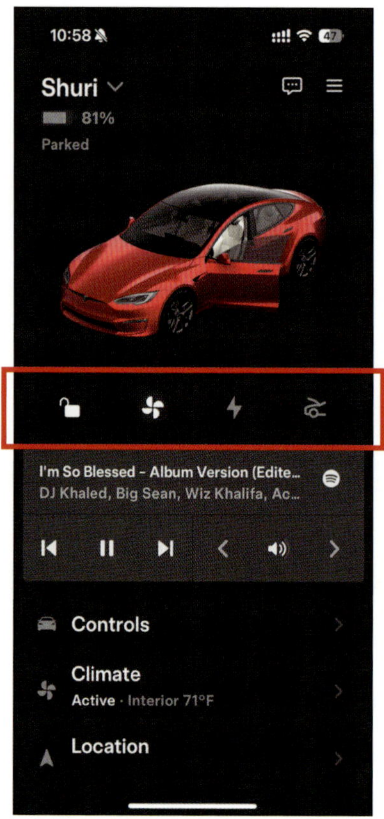

There are four Control (shortcut) buttons right beneath the graphic of your Tesla in the app. These are for doing things quickly without having to go into the Controls screen. By default, you'll see icons for locking/unlocking your Tesla, turning on/off the climate, opening the Charge Port door, and opening the front trunk. Just tap on any of these icons to perform those operations. If you want to change the default buttons that appear here (like I have), see the next page.

How Do I... Customize My Control Buttons?

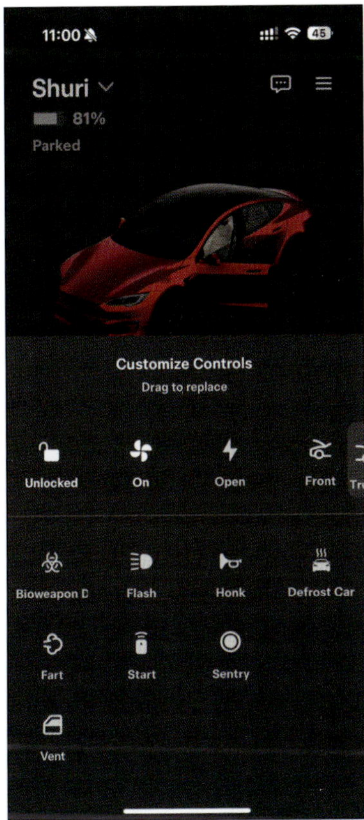

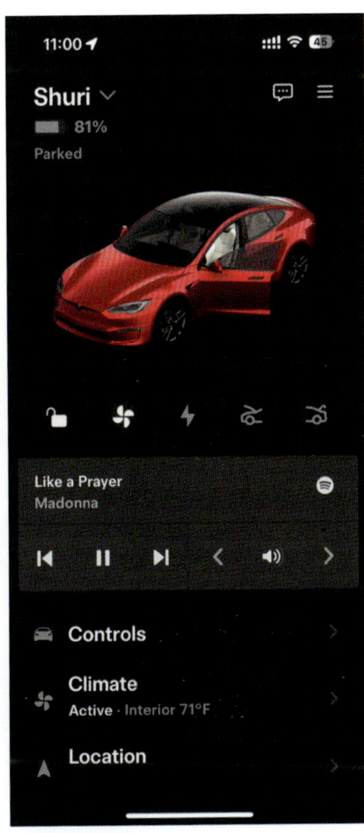

If you're like me, you access your trunk way more than you do your frunk (front trunk). So you may be asking why on earth did they pick that one to be there and not the trunk? Luckily, you can customize which buttons appear here. Just tap-and-hold on any of the four current buttons to bring up the Customize Controls screen. Here, you can tap-and-drag the ones you want to replace the ones you don't want. That's cool, but I've got a bonus tip for you: You can get a fifth Control button here. Just tap-and-drag the one you want over the fourth button and hold for a few seconds, then drag it to the right until you see an indicator that it's adding the fifth icon, and then let go. Now you have five Control buttons that can be customized or swapped out. *Note:* When you add the fifth button, you really need to drag the icon to the absolute edge of your screen and you might have to take off your phone case to do it.

How Do I... Sync My Calendar?

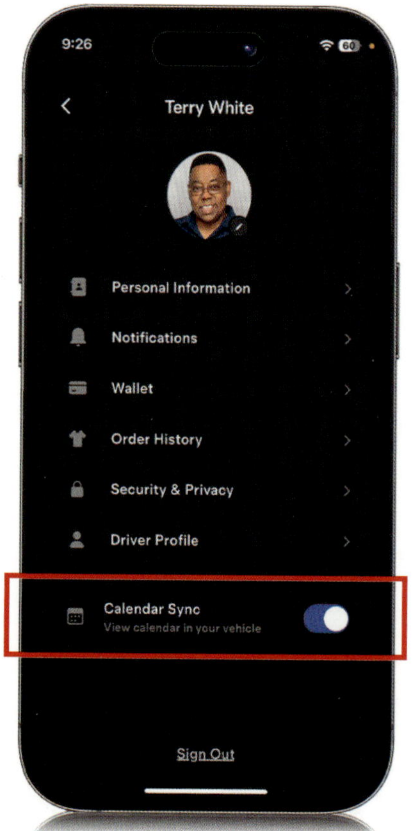

Your Tesla has built-in calendar support; however, it's not designed to really input appointments in your car. Instead, it's designed to sync with the calendar you use on your smartphone. To turn on Calendar Sync, tap the three-line (hamburger) menu icon in the upper-right corner of the Tesla app, and then tap on your profile (your name/email address). At the bottom of this screen, you'll see the option to turn on Calendar Sync. Your phone may ask for permission for the app to access your calendar, so if it does, grant it. The next time you get into your Tesla, you should be able to see your calendar in the Calendar app in the app tray at the bottom of your center touchscreen. Also, if there is an upcoming calendar appointment with an address in it, your Tesla should automatically start routing you there if it's near the same time as your appointment (see page 145 for more on this). That's pretty cool and saves you the step of having to enter it manually.

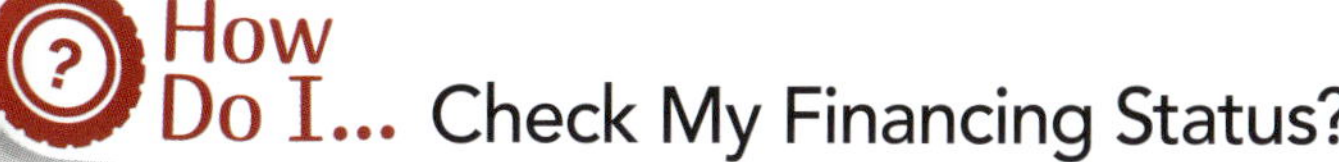

How Do I... Check My Financing Status?

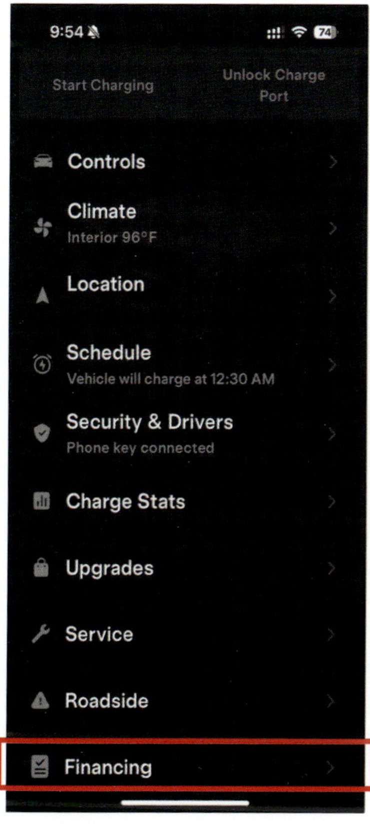

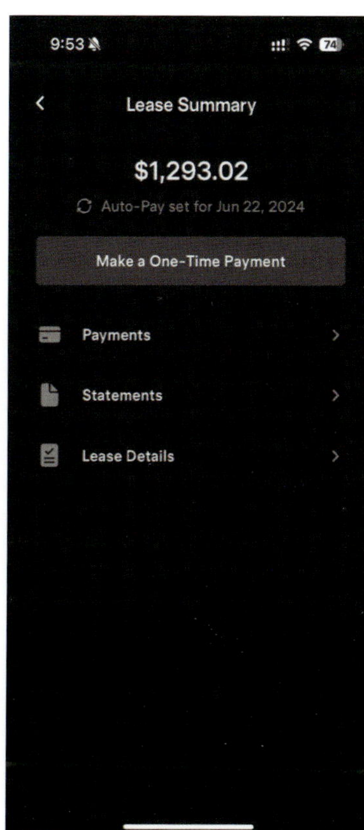

If you used Tesla financing to buy or lease your Tesla, you can see the details of your loan/lease agreement right in the Tesla app. Tap on Financing, and on the next screen, you can see your monthly payment, make a one-time payment, see the payments you've made, and see statements and details about your financing/lease contract.

How Do I... Buy Tesla Accessories?

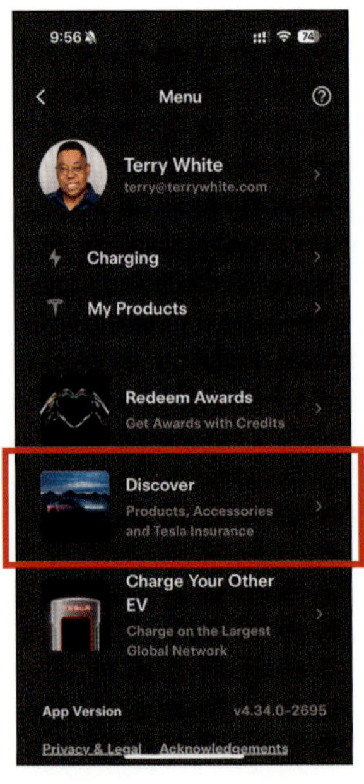

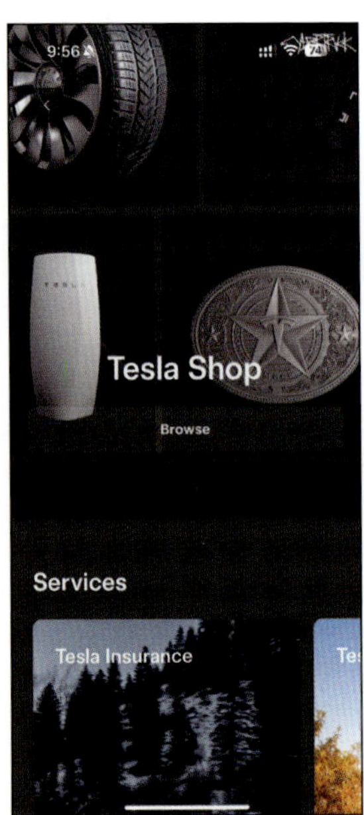

Besides all the cool things we've covered so far that you can use to control your Tesla from the Tesla app, you can also go shopping in the app. Tap the three-line (hamburger) menu icon in the upper-right corner of the app, and then tap on Discover. From this screen, you can go shopping at Tesla and buy everything from Tesla-branded clothing, to accessories for your vehicle, to Tesla Insurance (if it's offered in your location), to Tesla energy products, schedule test drives, and even order your next Tesla. When you tap the Browse button beneath Tesla Shop, you'll see the most common things people buy for their Teslas or even Tesla apparel. Anything you order will either be shipped to you or to your local service center if it requires installation.

How Do I... Use Siri Shortcuts to Voice Command My Tesla from My iPhone?

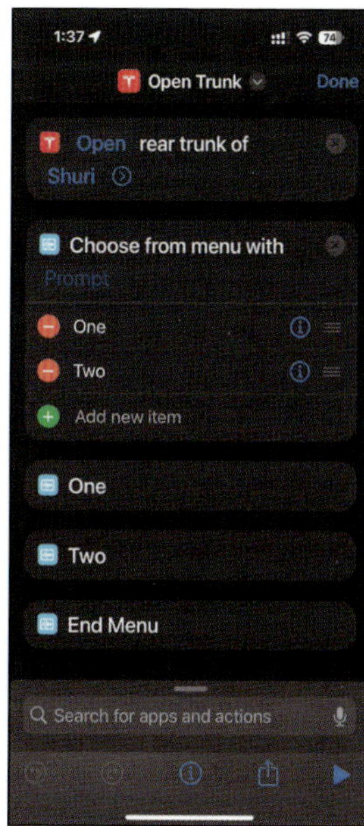

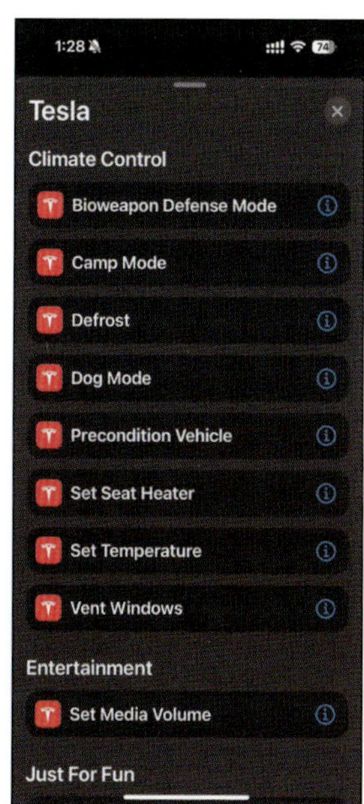

Built into every iPhone is the ability to use Siri, Apple's voice assistant, and Apple allows third-party developers to add Siri shortcuts to their iPhone apps. Tesla has done this, allowing you to control a variety of features right from your iPhone, or even your Apple Watch. You can say things, like "Hey, Siri, open the trunk" (that's my favorite one by the way). Although there are over 25 commands built in, you need to set them up in the Shortcuts app on your iPhone. So, launch the Shortcuts app (download it from the App Store if you deleted it), and then tap the + (plus sign) icon at the top right to create a new shortcut. Type "Tesla" in the search field to find the Tesla app, along with a list of Tesla commands, and then tap on it. Now scroll to the command that you would like to use—for example, "Close All Windows," or "Open Trunk"—and tap it. If your chosen command can do more than one operation, such as "Open or Close the Trunk," you'll need to choose which one you want, and if you have more than one Tesla, you'll need to choose the one you want this command to work with. If you want the voice command to be something unique, like "Open the back," instead of "Open Trunk," tap the downward-facing arrow at the top of the screen and rename your shortcut (this name is what Siri is looking for you to say), and then tap Done. Now you can invoke Siri on your iPhone or Apple Watch and say your new voice command to have the action happen on your Tesla.

How Do I… Add Tesla Widgets to My iPhone's Home Screen?

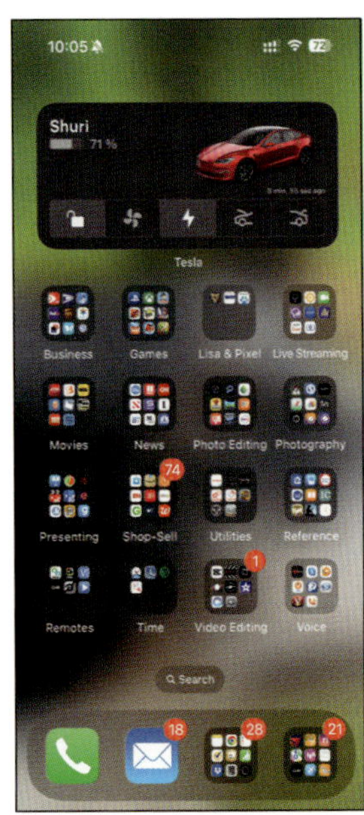

While it's great that you can see your Tesla's state of charge and even control it from the Tesla app, there's an even faster way, thanks to support of iOS widgets, because you can add a Tesla widget to your iPhone's Home Screen. Tap-and-hold on a blank area of your Home Screen and your apps on that screen will wiggle. Now tap on Edit in the top-left corner of your screen, then tap on Add Widget, and you will see a list of widgets available on your iPhone. Scroll down until you see Tesla, and then tap on its icon. Now you'll see a choice of four Tesla widgets: The first two are for your vehicle—there's a small one and a large one. The second two widgets are for your Tesla home energy products, like solar panels and Tesla Powerwalls. The smaller one shows your battery charge and allows you to unlock/lock your Tesla, while the larger one has those features plus the rest of your Control shortcuts. Once you make your size choices, you can move the widget around from Home Screen to Home Screen like you do any other app or widget. You can even add different versions to different Home Screens. Last but not least, you can also add a Tesla widget to your iPhone's Lock Screen to see your Tesla's status, even if your iPhone is locked. To add a Lock Screen widget, just tap-and-hold on your Lock Screen and then tap the Customize button to customize the widgets on your Lock Screen. You can add up to four widgets on your Lock Screen.

How Do I... Add Tesla Widgets to My Android's Home Screen?

Add a Tesla widget to your Android's Home Screen by tapping-and-holding on a blank area of your Home Screen and your apps on that screen will wiggle. Now tap the Widgets icon at the bottom of your screen and you will see a list of widgets available on your Android. Scroll down until you see Tesla and tap on its icon. Now you'll see a choice of four Tesla widgets: Tap on either the first or the second one, depending on your preference in size. The smaller one shows your battery charge and allows you to unlock/lock your Tesla, while the larger one has those features plus the rest of your Control shortcuts. From this point on you can move the widget around from Home Screen to Home Screen like you do any other app or widget. You can even add different versions to different Home Screens.

Chapter 8

How to Use Autopilot & Full Self-Driving

The Future of Driving. Or Not

I love to drive. Always have. I'm like that 16-year-old who just got their driver's license, and for this one short sliver of time in their life, they are willing to run any errand for their parents just for an excuse to get in their car and drive again. So, you might ask why I am at all interested in having my car drive around for me? It's because other drivers are awful drivers. Not me, mind you. I'm solid. But, other folks. Crazy. I often yell at them while driving, but it's much easier to yell and perhaps add hand gestures if the car is driving for you. We want the freedom to curse and gesture without worrying about driving off into a ditch. Why is this? It's simple. Drivers, by nature, are crazy. I don't mean like, "Oh man, that guy is crazy!" type of crazy. I mean the "Uh-oh. That guy is crazy" kind of crazy, which is an entirely different type of crazy, and which 96.3% of all drivers (except myself, of course, and Terry) qualify as. The other 3.7% are just nuts. So, this begs the question: "Are any people out there driving around not crazy?" Well, for the sake of drama, I was exaggerating a bit when I said 96.3% of all drivers are crazy. In reality, it's really more like just 94.6%, and that's according to the latest statistics from the Automotive Statistics Society (or the ASS, for short). In all honesty, I hadn't heard that much about the ASS, and I wanted to make sure it was a legit organization, so I did a little digging around inside the ASS, and what I found in there was remarkable. There's loads of stuff in there—a lot more depth than I expected—so I really started diving in. Now, it's not all roses and sunshine in there. The deeper you look, the darker it gets, but still, I applaud the whole ASS, and I was especially impressed with their astrophysics initiative and their study of Uranus. Okay, wait! Even I have to speak up at this point. Why does my publisher, an otherwise respectable publishing organization, allow me to get away with this type of stuff? I'll tell you why—they don't read anything I write. Ever. They publish it, sight unseen, and hope for the best. If they did read it, surely they would report me to the ASS, but they don't want to go down that rabbit hole, now do they? See? Not reading. Not a word.

How Do I... Use Autopilot?

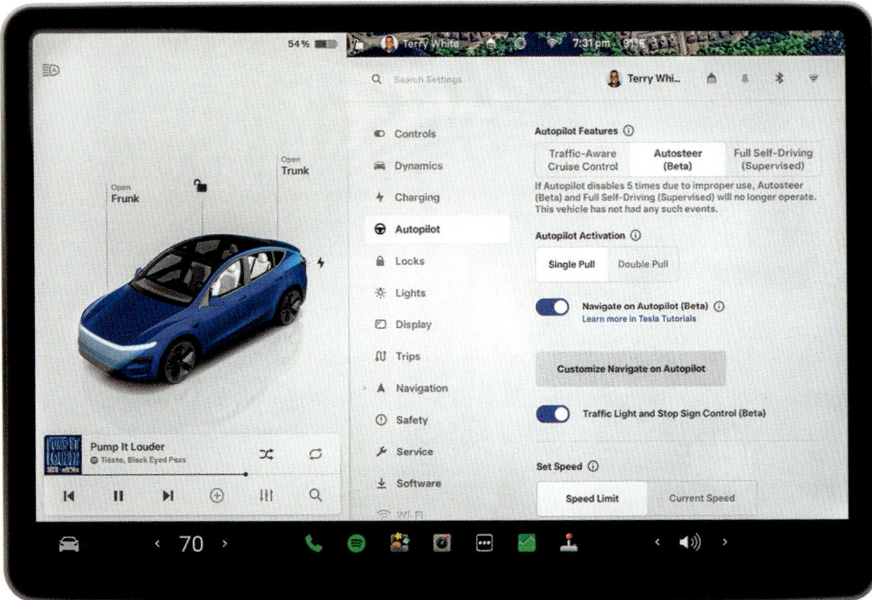

Every Tesla sold today (since September 2014) comes with Autopilot. You don't pay extra for it (anymore); it's a standard feature. You get two features with it: The first is Traffic-Aware Cruise Control, which is for highway driving, and it simply matches the speed of the vehicles in your surroundings up to a set speed limit. If the vehicle in front of you slows down or stops, your Tesla will slow down and stop if needed. You are responsible for steering and staying in your lane. The second feature is Autosteer, which will steer your Tesla on highways with clearly marked lanes, and also uses Traffic-Aware Cruise Control. It will *not* make turns for you. It will not stop at stoplights and stop signs. It's basically a form of cruise control that stays in a clearly marked lane. You'll need to stay alert and, occasionally, apply light torque to the steering wheel to let it know that you are awake and paying attention. See page 206 for more about Autopilot's settings.

How Do I... Use Enhanced Autopilot?

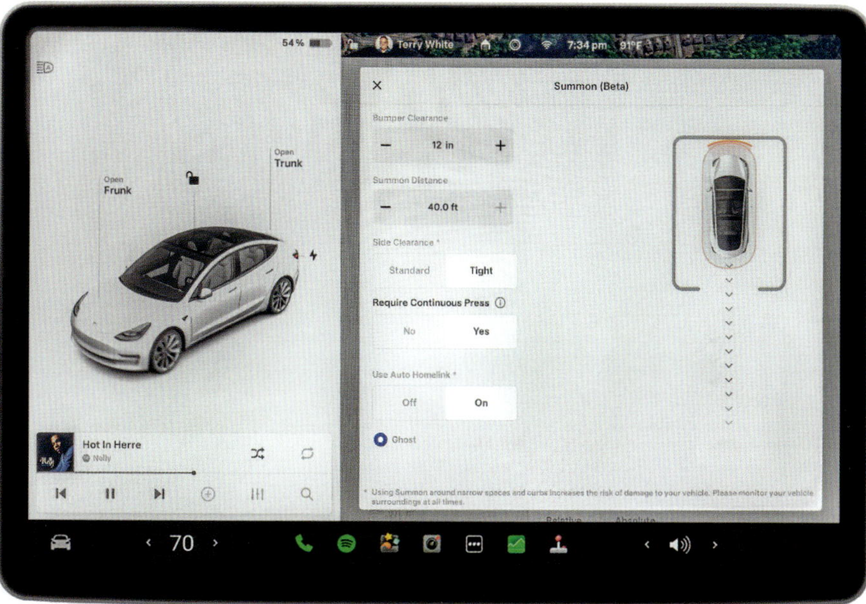

This feature is no longer available; however, I'm covering it here just as an FYI for those that bought a Tesla that included it or paid extra for it. As you may have guessed, Enhanced Autopilot is a version of Autopilot that fit in between Autopilot and Full Self-Driving (Supervised). When I got my Tesla in 2023, I bought the Enhanced Autopilot option, which includes all the features of Autopilot (seen on the previous page), along with Auto Lane Change, which can move your Tesla over to the next lane by engaging your turn signal while on the highway. It also includes Navigate on Autopilot, which automatically performs lane changes as needed to navigate to the destination in your Tesla's navigation system from the on-ramp to the off-ramp. Lastly, Enhanced Autopilot includes Autopark and supports Summon features (I cover these in detail later in this chapter). Again, Tesla replaced Enhanced Autopilot with Full Self-Driving (Supervised), but existing Teslas with it did not automatically switch over to FSD. Owners would need to choose this upgrade.

How Do I... Use Full Self-Driving (Supervised)?

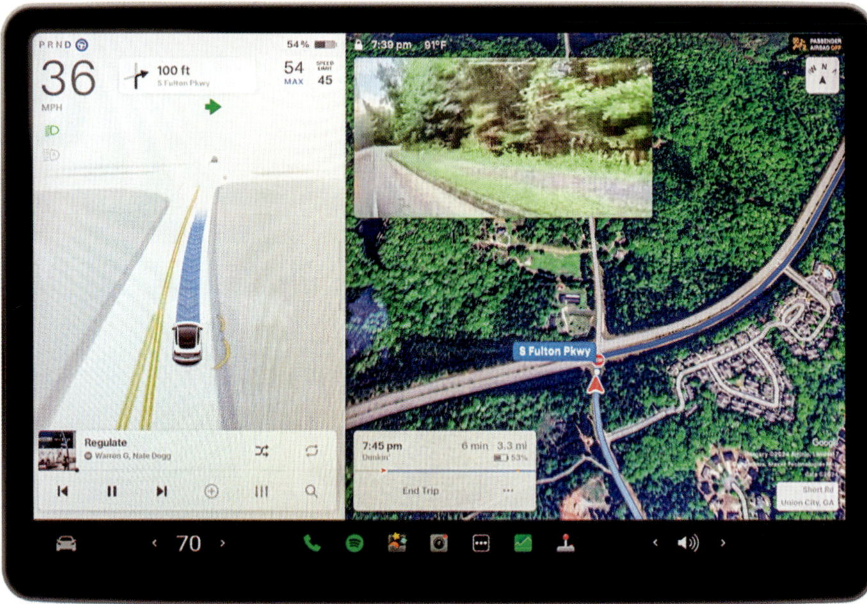

Tesla has been working on autonomous driving for years and we're finally getting to the point where it actually works. It is by no means perfect and makes mistakes all the time, so you have to stay alert, pay attention, and be ready to take over at any time. I don't expect to be able to jump into my back seat and take a nap anytime in the near future. In addition to all the features of Autopilot and Enhanced Autopilot (covered on the previous pages), Full Self-Driving (Supervised) (FSD) includes Autosteer on City Streets where it can make turns to follow your route, and Traffic Light and Stop Sign Control, which identifies stop signs and traffic lights. It will automatically slow your Tesla to a stop, and when the traffic clears at a stop sign or the traffic signal turns green, your Tesla will automatically start moving again. FSD has an additional cost, and as of the writing of this book, it is $8,000, but you can do a no contract/cancel-at-anytime monthly subscription for $99/month. Tesla has recently rebranded this feature from Full Self-Driving (Beta) to Full Self-Driving (Supervised). Although they removed the word "Beta" from their branding, I still very much consider it to be a work-in-progress feature and it should be used with caution.

How Do I… Subscribe to Full Self-Driving (Supervised)?

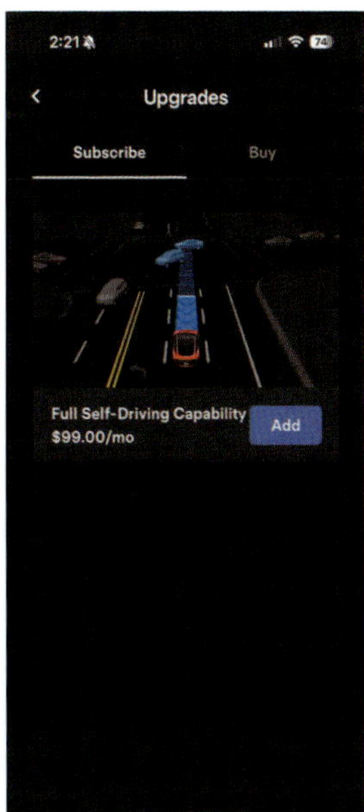

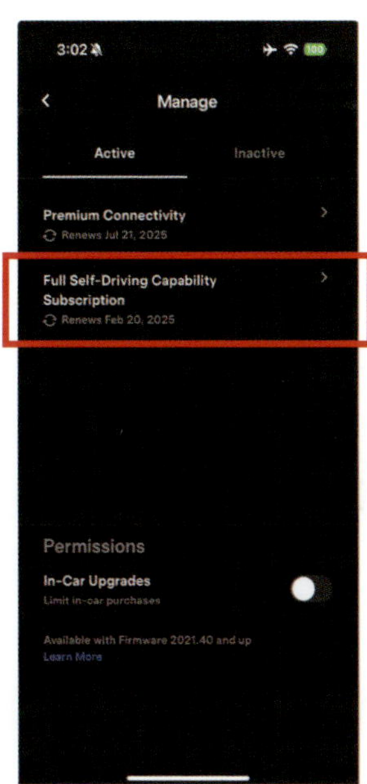

I love Full Self-Driving (Supervised) (FSD) and use it daily. However, I would be hard pressed to recommend buying it for $8,000, let alone when it used to go for as high as $15,000! I know that people hate subscriptions, but now that Tesla has lowered the FSD subscription to $99/month (as of the writing of this book), down from $199/month, it's a much easier recommendation. There is no contract for the FSD subscription and no fee for canceling. You can turn it on or off as needed. If you want to use it for a road trip, but don't need it day to day, you can subscribe for one month and then cancel it. You should do the math, though. If you plan to keep your Tesla for several years, then paying the one-time, up-front cost of $8,000 may be cheaper, but that's a big if. If you decide to buy another Tesla, FSD is not transferable (again, as of the writing of this book). Sometimes Tesla does run deals that allow you to transfer FSD to a new vehicle, but that is not a guarantee. To subscribe to FSD, open your Tesla mobile app, tap Upgrades, and under Subscribe, you should see an option to add Full Self-Driving Capability for $99/month (or the current rate, if it has changed). If at any time you want to cancel your subscription, tap Upgrades, and then tap Manage Upgrades at the bottom. Tap Full Self-Driving Capability Subscription, and you'll see a Cancel Subscription button. Just tap it to cancel your monthly charge. It should work for the remainder of the paid time you still have, but it will not renew.

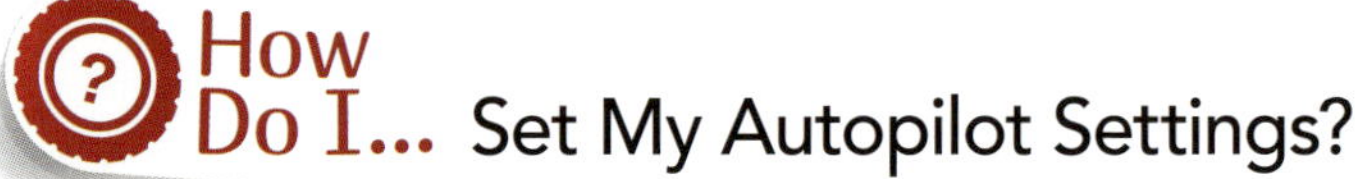

How Do I... Set My Autopilot Settings?

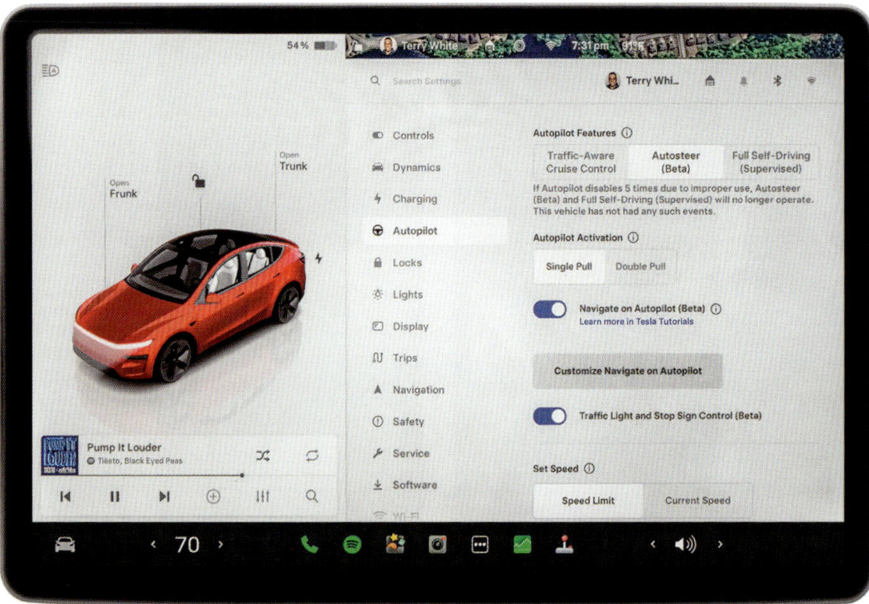

Autopilot is included with your Tesla, but you will need to configure your settings before you can use it. On your touchscreen, tap Controls (the car icon), then tap Autopilot, and then tap Autosteer (Beta). Beneath Autopilot Activation, choose how you'll activate your Autopilot. By default, a single pull of the drive stalk or a single press of the right scroll button on your steering wheel (if your Tesla doesn't have stalks) will activate Traffic-Aware Cruise Control and a double pull (or double press) will activate Autopilot (Autosteer). If you don't ever use Traffic-Aware Cruise Control, then you can set Autopilot to activate with a single pull or single press (see the next page for more on this). Next is Set Speed, where you can choose whether Autopilot starts out at the current speed limit or your current speed. Further down this screen is one of my favorite settings and that is Offset. Most people drive faster than the speed limit, so if that's you, set a Speed Offset using Fixed (which adjusts the speed by a specific amount) or Percentage (which is a percentage of the road's detected speed limit). The last setting, near the bottom of this screen, is Green Traffic Light Chime, which is a nice feature that simply plays a chime when the light turns green or the car in front of you moves forward. (*Note:* This feature is currently only available in the US and Canada.)

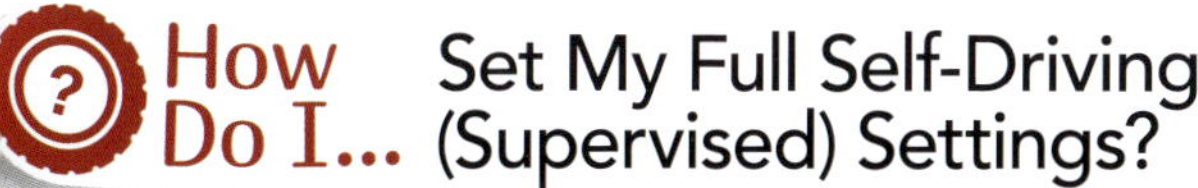

How Do I... Set My Full Self-Driving (Supervised) Settings?

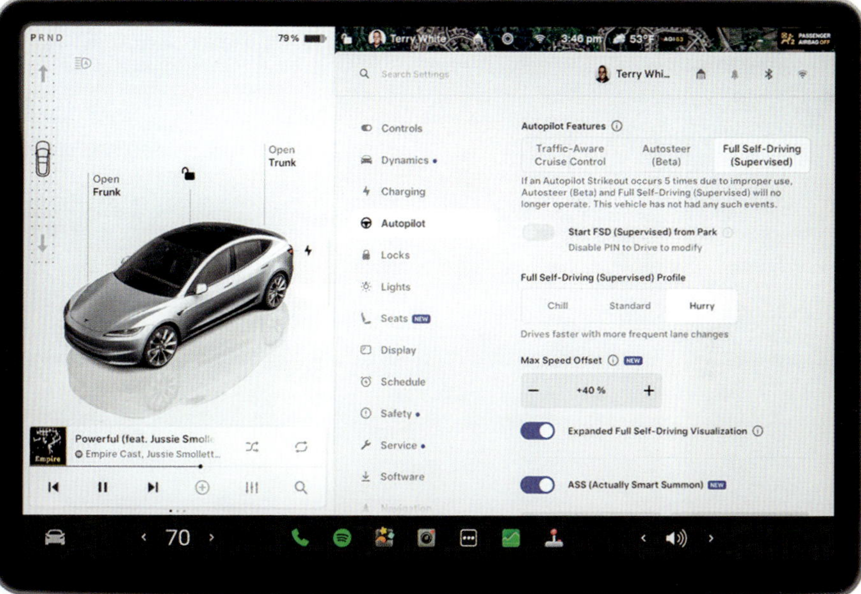

If you purchased Full Self-Driving (Supervised) (FSD) or subscribed to it (see page 205 for more on this), you will need to enable it in the settings first before you can use it. On your touchscreen, tap Controls (the car icon), then tap Autopilot, and then tap Full Self-Driving (Supervised). If this is your first time using it, you'll likely get a pop-up with the terms and conditions that you should read carefully before accepting. The first thing you should note is that with FSD enabled, you will no longer have access to Traffic-Aware Cruise Control, and you'll activate FSD with a single pull of the drive stalk or a single press of the right scroll button on your steering wheel (if your Tesla doesn't have stalks). Now you'll want to choose between the three profiles: Chill, Standard, or Hurry. As the name implies, Chill is a more relaxed driving style and Hurry drives with more urgency, with Standard falling somewhere in between. Until you get used to FSD, I recommend starting out on Chill. There is an option called "Minimal Lane Changes for Current Drive," and it prevents your Tesla from making a bunch of lane changes on the highway, which could be frustrating in heavy traffic. If you enable this feature, it will only be used for the current drive and will switch off for the next one (I truly wish that Tesla would move this option to the main touchscreen, so that you could tap it during a drive without having to go into this screen). Finally, there is the option to enable Automatic Set Speed Offset. When enabled, your Tesla will drive at a speed that Autopilot determines to be the "most natural." I don't like this feature because, for me, it always drives too slow. However, give it a try; you may prefer it.

How Do I... Engage Full Self-Driving (Supervised)/Autopilot?

The first thing you should do is put your destination into your Tesla's navigation system, even if you already know the way because FSD has no idea where you're going. If you don't put in a destination, it will still drive, but it won't make turns because it doesn't know where it's going. Now that you've entered your destination, look to see if you have the Autosteer steering wheel icon on your center touchscreen. This indicates that you're on a road where you can use FSD. For example, you may not be able to start FSD while you're in your garage or in a parking lot—you may have to pull out onto the street first. Once you see the Autosteer steering wheel icon, this means that you can enable FSD. You'll need to be in Drive, as it won't shift out of Park for safety reasons. Then, you can pull the drive stalk once (or twice depending on your settings or Tesla model; see page 207) or press the right scroll button on your steering wheel once (if your Tesla doesn't have any stalks). You'll hear a chime confirming that you're now using FSD and you should see a blue trail in front of your Tesla on your touchscreen. Remember that at any time, you can take over (intervene) by pressing the brake pedal, pushing your stalk up once (or pressing the right scroll button), or by turning the steering wheel.

How Do I... Know If Full Self-Driving (Supervised) Knows I'm Paying Attention?

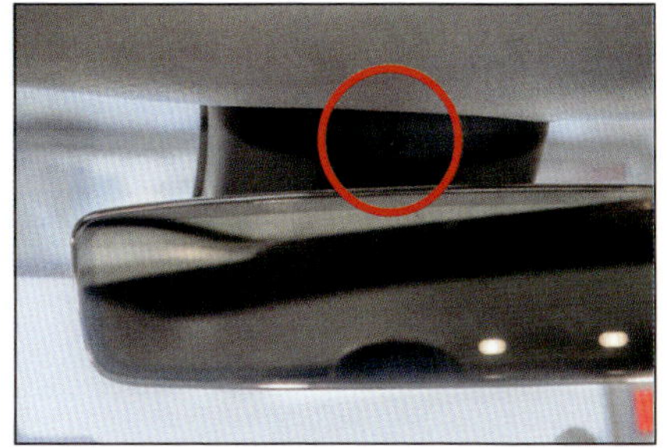

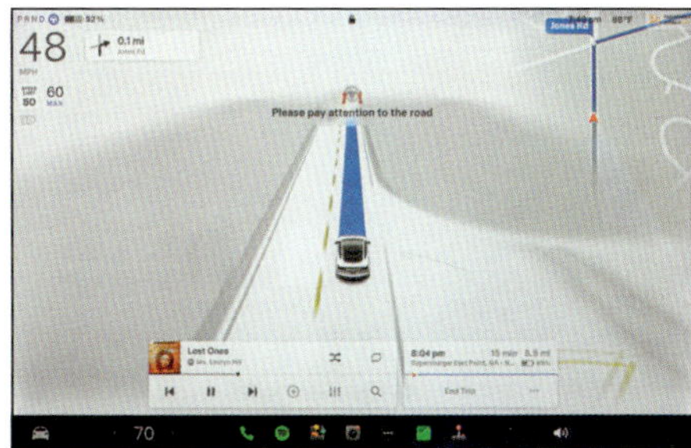

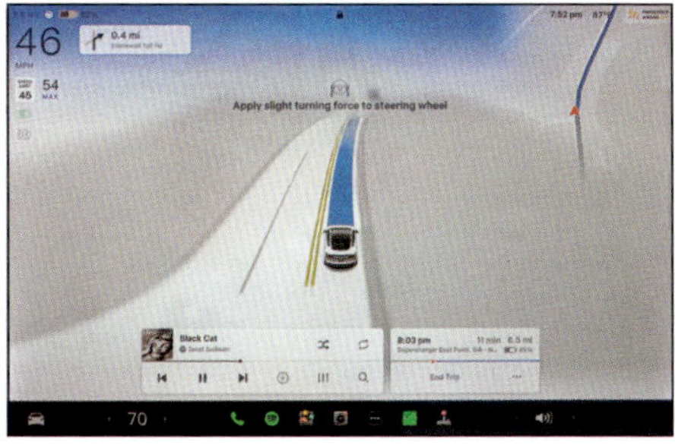

It's important that you pay attention and stay alert while using Full Self-Driving (Supervised) and you have to be ready to take over at any time. There are two methods that your Tesla uses to make sure that you're alert and paying attention: The first and oldest method is that you have to regularly apply a little torque to the steering wheel, and simply holding it is not enough. You have to ever so slightly apply torque as if you're going to turn the steering wheel without actually turning it, as that would disengage FSD. If you go too long without applying torque, you'll be notified via your touchscreen and if you ignore it, FSD will be automatically disengaged and a strike will go against you (see the next page for more on this). The second way that Tesla monitors whether or not you're paying attention is via the cabin camera located above the rearview mirror. This camera constantly monitors whether or not you're looking forward at the road. If you're looking down (like checking your phone) or away from the road for too long, you'll start to be notified on your touchscreen, and eventually, FSD will disengage for the remainder of your trip. If a forced disengagement happens, you won't be able to re-engage FSD unless you pull over, park, exit the vehicle, close the door, and then get back in again. In other words, until you start another trip. As of the writing of this book, Tesla rolled out an FSD software update, which eliminates the steering wheel nag, as long as your Tesla has a cabin camera and you're not wearing sunglasses, a hat, or anything that blocks the camera from seeing your eyes.

How Do I... Know What "Five Strikes You're Out" Means?

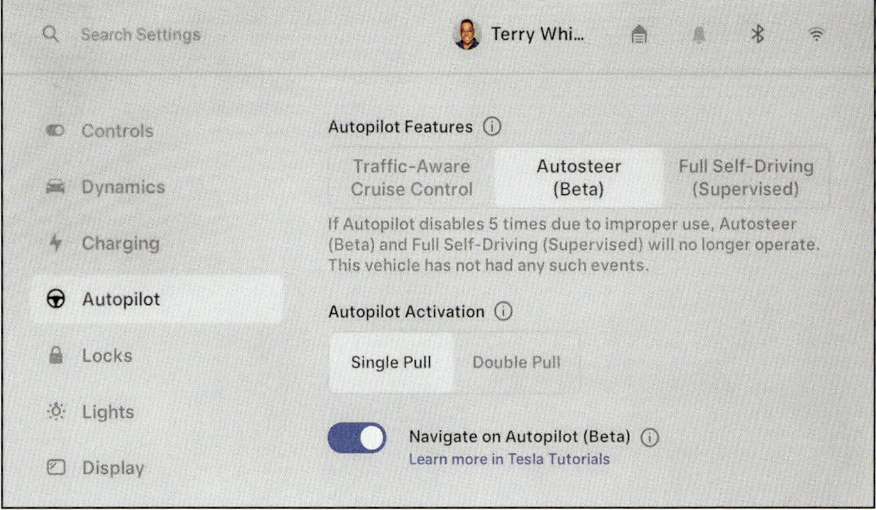

If you break the rules and rack up five forced disengagements, you could lose access to Full Self-Driving (Supervised) for a week. A forced disengagement is when the Autopilot system disengages for the remainder of a trip after the driver receives several audio and visual warnings for inattentiveness. In other words, you were a distracted driver and ignored the warnings (see the previous page for more on this). In all the years of using FSD (Beta) and (Supervised), I've only had one forced disengagement and it was my fault—I kept looking at my smartphone after ignoring several warnings. That was a wake-up call for me and so far, I haven't let it happen again. Now, if you're thinking, "Hey, I paid for this. You can't take it away from me!" Think again. You agreed to these terms when you activated Autopilot/FSD (Supervised). Keep your eyes on the road, don't drive more than 85mph while on FSD, and you'll be fine. Note that there is no strike, if *you* take over (see the next page for how to take over). This is only if FSD forces a disengagement.

How Do I... Take Over from FSD/Autopilot in an Emergency?

Tesla Autopilot and Full Self-Driving (Supervised) are far from perfect, and like any AI, they make mistakes, and these mistakes could cause an accident. That's why it's important that you stay alert and ready to take over at any time. If your Tesla is about to do something that endangers you or others, you should take over immediately—don't second guess yourself as it's better to be safe than sorry. Luckily, the software has gotten really good, but I still have to take over occasionally. There are basically three ways to take over in an emergency: The first is more of a nudge than an actual takeover. Sometimes, when my Tesla stops at a stop sign, it will just sit there, making sure that it can make the turn, that I'm looking both ways, and there isn't a car coming. So, I'll press the accelerator to make it go instead of continuing to let it evaluate if it's safe to do so. Same thing if I'm on the road and it's not going fast enough—I'll press the accelerator to speed it up. Neither of those are dangerous to anyone, but it's nice to have the option to make it go or go faster without actually taking over. The second way is to simply hit the brake. This completely disengages FSD and slows/stops your Tesla. It reminds me of when I took driver's training and the car was equipped with a brake pedal on the passenger side for the instructor to be able to stop the car in an emergency. Same thing—you can stop your Tesla whenever needed. The third way is by turning the steering wheel. I rarely ever have to grab the wheel, but like hitting the brake, it will disengage FSD immediately (it won't stop your Tesla unless you also press the brake pedal). Like I said earlier, I use FSD every day and as good as it has gotten, there isn't a week that goes by that I don't have to take over at least once.

How Do I... Report Errors and Mistakes?

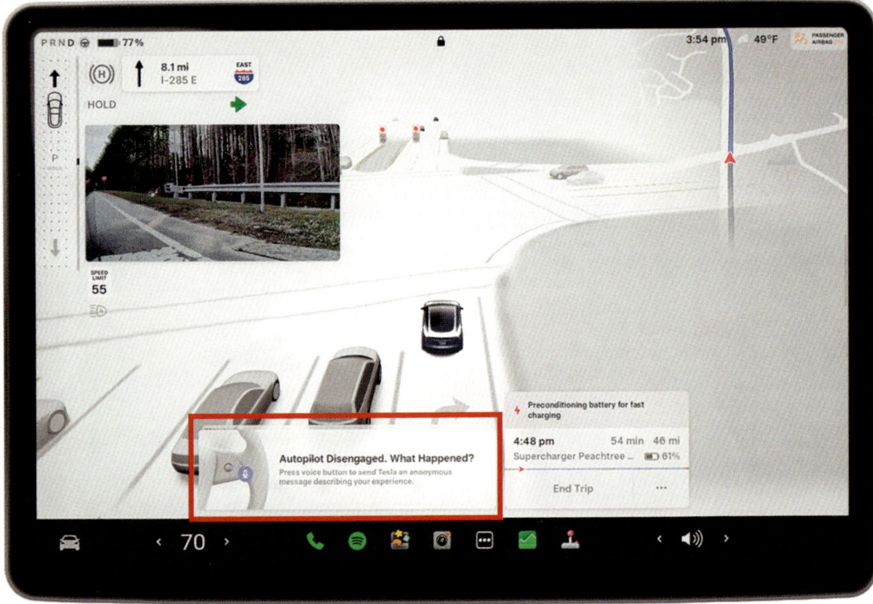

Full Self-Driving (Supervised) is still very much a work in progress and it makes mistakes that sometimes require you to take over (see the previous page for how to take over). You'll notice when you take over that a message pops up on your touchscreen asking why you disengaged, with the option to use voice command to tell Tesla about your experience. Press the Microphone button on the right side of your steering wheel (or the right scroll button in older models), then release it and say why you had to disengage. This feedback goes to the Tesla FSD development team so that they not only have the data from your Tesla, but they also have your verbal explanation. It's anyone's guess as to how many of these recordings actually get listened to, but at least they give us the option to report problems. Unless I'm on a phone call, I always report these errors. Once you're done speaking, press the Microphone button again to send the feedback to Tesla.

How Do I... Know How Bad Weather Conditions May Impact Using FSD (Supervised)?

©VICTORIA WHITE

Tesla's Full Self-Driving (Supervised) operates by using an array of cameras located around your Tesla. If those cameras are obstructed by dirt, rain, snow, fog, or other conditions that are hard to see through, then you may see a message stating that Full Self-Driving (Supervised) is degraded. It may still work sporadically or it may not allow you to engage it at all—if it can't see, it can't drive. If this happens, you should take over and drive carefully. Also, in addition to keeping your external cameras clean, you should also keep your windshield clean as there is an array of cameras in your cabin behind the rearview mirror, facing outward through the windshield. A general rule to live by is that if it's hard for you to see the road, then it's really hard for FSD to see the road.

How Do I... Make Full Self-Driving (Supervised) Full-Screen?

Normally, you see the Full Self-Driving (Supervised) experience on the left side of your touchscreen and the map on the right side, with the map taking up the major portion of your screen. But you can enjoy a full-screen, Full Self-Driving (Supervised) experience where the FSD user interface takes up the whole touchscreen, putting the map in the upper-right corner (as seen here). You can enable this view by simply tapping-and-holding on the right edge of the FSD part of the screen (there's a visible tab there) and pulling it or swiping it to the right to make it take up the whole touchscreen. You can always go back to the split view by swiping the right edge to the left. If you always want this view when you engage FSD, tap Controls (the car icon) on your touchscreen, then tap Autopilot, and then turn on Expanded Full Self-Driving Visualization.

How Do I... Know What Will Happen When I Arrive at My Destination via FSD?

If you used Full Self-Driving (Supervised) (FSD) to drive to a destination, you may be wondering what happens when you arrive. In a perfect world, your Tesla would Autopark for you and that would be it (see the next page for more about Autopark). However, as of the writing of this book, FSD doesn't engage Autopark or disengage when you arrive. You'll need to manually disengage and take over, so just step on the brake and then take it from there on what you want to do next, including pulling into your driveway or parking.

How Do I... Use Autopark?

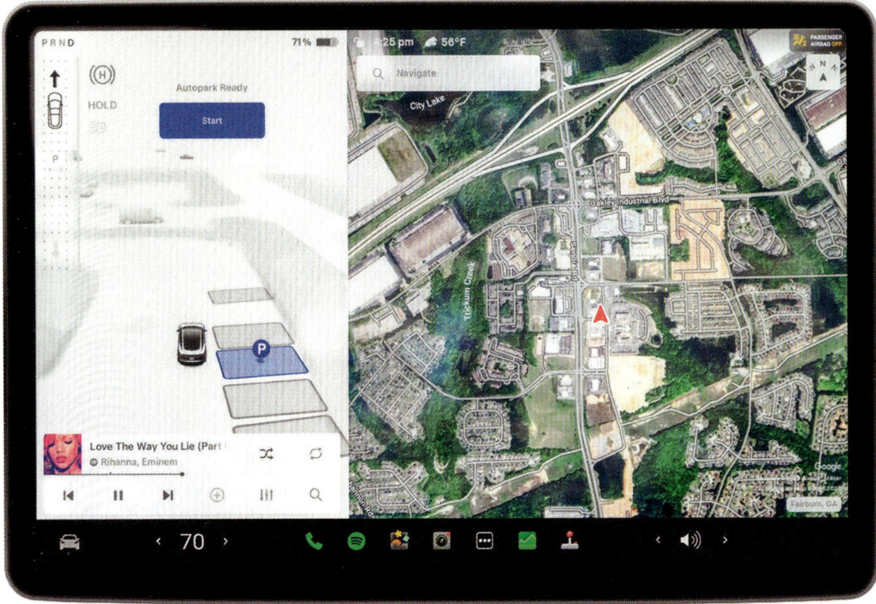

One of the benefits of Full Self-Driving (Supervised) is that you get the Autopark feature. Autopark can automatically park your Tesla in parking lots, as well as parallel park on the street. To Autopark in a space in a parking lot, drive slowly near the space(s) that you want to park in and Autopark should automatically detect the empty spaces. Once it detects the space that you want to use, press the brake to stop, then tap the Start button (under Autopark Ready) on your touchscreen, and then take your hands off the wheel and foot off the brake. (*Note:* On a Model 3 or Model Y, you can actually tap on the screen on the space you want.) Autopark always backs into the space in a parking lot, therefore, it may need to pull forward first in order to line up to back into the space. Parallel parking is very similar. Slow down as you approach the space you want to park in, and wait for Autopark to detect the space. Stop your vehicle, and then tap Start. Your Tesla should move forward, if necessary, and then begin backing into the parking space. It could take a couple of moves to get it parked, but it usually does an awesome job of ending up in the space perfectly.

How Do I... Use Dumb Summon and Actually Smart Summon (ASS)?

The first thing to get right is the name of this feature. I hear so many people refer to it as "summons," which is something you get requiring you to appear in court. Dumb Summon and Actually Smart Summon (ASS) are features that allow you to move your Tesla using the Tesla app. This sounds cool and it is, but let me set expectations first. Dumb Summon allows you to move your Tesla forward or backward, which is great when you need to move out of/into tight spots without being in it—I use it sometimes to pull out of my garage. Actually Smart Summon allows you to have your Tesla come to you in a parking lot. You will need to be able to see your Tesla to make sure it doesn't hit anything or anyone while it's coming to you. It's amazing when it works, and quite embarrassing when it messes up. To use Summon, open the Tesla app and tap Summon. You'll have three choices: forward and backward (the arrows) and Go to Target. For Dumb Summon, tap-and-hold the arrows to move your Tesla forward/backward. Actually Smart Summon uses the Go to Target button, which you tap-and-hold to have the Tesla come to you. If for any reason you need your Tesla to stop, simply release the button.

Chapter 9

How to Use Superchargers

More to It Than It Appears

Sooner or later you're going to find yourself at a Tesla Supercharger, which, if you have a charger at your house, will probably be on a road trip, but there are some folks who live in apartments or in prison and have to use Superchargers to charge their batteries (probably once a week or so). So, let's say you're on a road trip, and you stop at a Supercharger so you have enough battery to make it to your destination. The first thing you'll notice is that most Superchargers are well-named because they charge really super-fast. I mean crazy fast compared to the speed your Tesla charges at your house on a 240v dryer outlet. Now, I know what you're thinking: "Hey, I'm already using a 240v outlet, which is the maximum amount of electricity you can even have in a house. I'm sure the electricity for the Supercharger comes from the same power plant I get my house electricity from (in my area, it comes from Duke Energy, which I believe is owned by Prince William, Duke of Cambridge, Cornwall, and Rothesay). So, why doesn't my Tesla charger at home charge this fast?" Well, I'm glad you asked. The advantages Superchargers have is that they are wired directly to the grid, but it's not the standard power grid homes and businesses use. It's a special fission power grid, which is why you never see a Supercharger with more than 12 charging stalls. If they add more than 12, there's a distinct chance (greater than 68%) that the fuel rods would start to melt, which could breach the fuel cladding and cause a chain reaction that could potentially release radioactive materials into the air surrounding the Supercharger (most likely a Wawa or Target parking lot), which would then be contaminated and unusable until the radiation release subsides, which would be sometime in the year 2082, if all goes well. But, it never does in situations like this, so you can forget about going to that Target for your holiday shopping for a good long while. Now you're having to schlep all the way over to that Supercharger by the interstate, and that one's always full since the one in the Target parking lot "melted down." So, I guess what I'm trying to say here is that while it's slower to charge your Tesla at home, at least your kids and pets won't be glowing.

How Do I... Plan for Charging on My Trip?

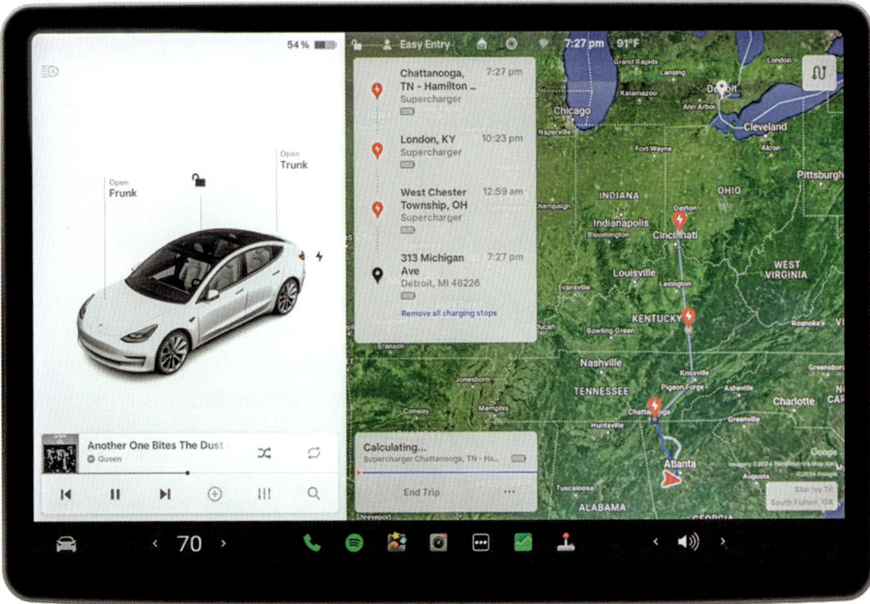

Most Tesla owners charge regularly at home or at work and only need to use Superchargers on road trips. Supercharger rates are typically much higher than your home electric rates (unless you live in California where all bets are off). If you only use Superchargers on long road trips or when you forget to charge at home/work, then you should start by putting your destination into your Tesla's navigation system. It will then figure out if you need to charge to make it to your destination, the closest Supercharger(s) along your route, and how long you'll need to charge. This takes the work out of planning your trip. As you get close to a Supercharger, it will even display how many stalls are available or if there is a wait, and you'll also see what the rates are for that location.

How Do I... Know How Much to Charge? (Hint: It's Not 100%)

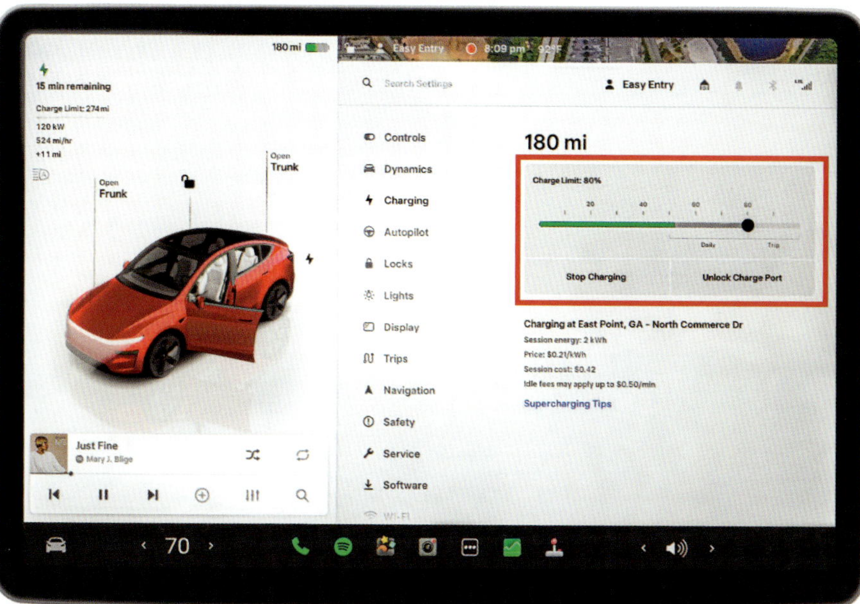

Unlike a gas vehicle, where you would normally fill the tank with gas at each stop, EVs are different—you only need to charge enough to make it to the next stop or your destination. You can charge to a higher state of charge, but there's a big disadvantage of charging over 80%–90%. Think of your Tesla's battery as if it were a big empty auditorium. When you open the doors and people pour in, they can find seats pretty quickly. However, as the room fills up, it takes longer to find a seat. Charging your Tesla works the same way. When your state of charge is low, you'll notice that your Supercharger speed is really fast. However, as your battery goes past 80%, the charge rate slows down considerably. It could take as long to go from 80% to 100% as it did to go from 20% to 80%. That's why it's best to make more shorter stops than fewer longer stops. Plus, it also frees up the Supercharger for the next person.

How Do I... Know How Long It Will Take to Charge?

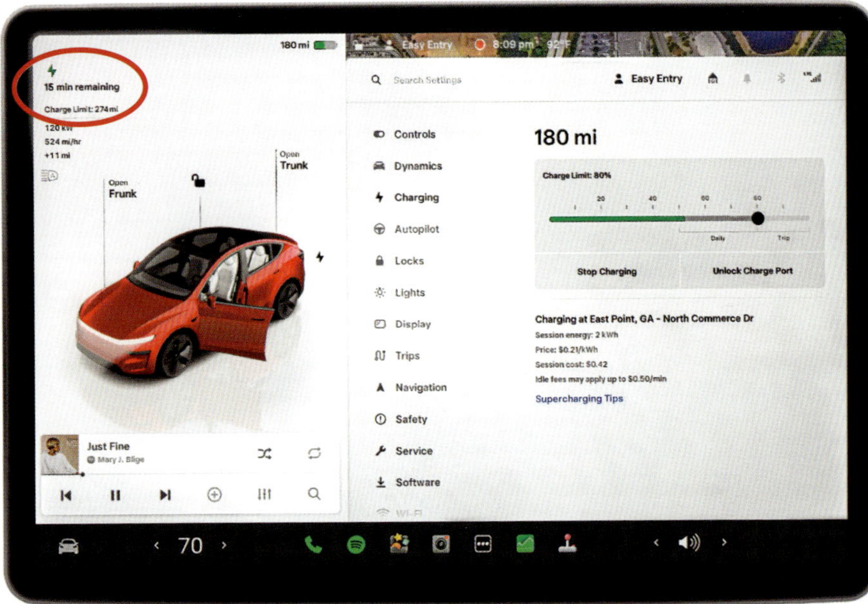

When you plug your Tesla into a Supercharger, you'll see an estimate on your touchscreen of how long it will take to reach your charge limit (see page 107 for more on setting this limit). If you used the Tesla navigation system to get to the Supercharger during the road trip, there was also an estimate on your touchscreen of how long you would need to charge before you could continue on. You can always charge longer to give yourself a cushion, but that's up to you and your schedule. As I've mentioned, your Tesla will charge faster if your current state of charge is low (0%–50%) because as your battery fills up, the charging rate will drop and slow down. This is normal and just the way batteries work. Most Supercharger stops are 15–20 minutes.

How Do I... Know the Different Supercharger Speeds?

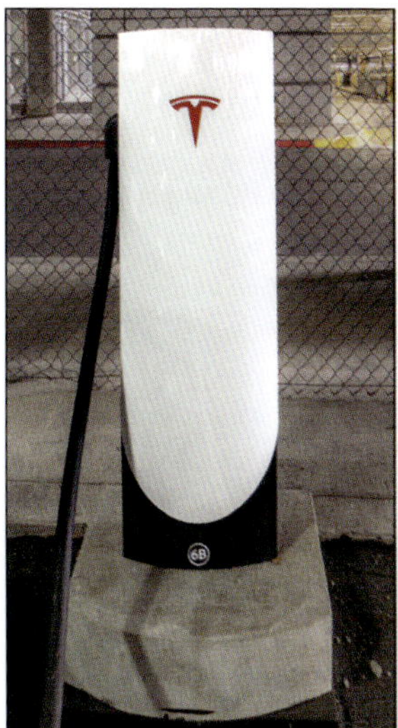

Urban Supercharger 72 kW

Supercharger V2 150 kW and V3 250 kW

Supercharger V4 250 kW. Also has the CCS Magic Dock to charge other EV brands

Not all Superchargers are created equal. The older version 2 Superchargers charged at a rate of 150 kW. This was considered fast until the version 3 and 4 Superchargers were released, which charge at a rate of 250 kW. The version 3 and 4 Superchargers also don't share power between stalls, so therefore, all the Teslas will charge at the fastest rates available. There is another Supercharger model, which is normally found in parking lots of shopping centers, and these smaller Urban Superchargers only charge at 72 kW. Yes, that's much slower, but these are rarely located in places where people are doing road trips. So, if you're at the mall shopping, these chargers are usually fast enough to give you a significant charge much faster than the 240v Level 2 charger you may have at home.

TIP: FIND NEW SUPERCHARGER LOCATIONS

If you're interested in seeing the progress of new Supercharger locations that are opening, here's a great site for that: https://supercharge.info/map.

How Do I... Know If a Supercharger Is in a Pay-for-Parking Lot?

Most Superchargers are in easily accessible lots where you just pull in and charge. However, at some locations, the Supercharger is located in a parking lot that you would normally have to pay to use. Fortunately, Tesla usually works out deals with these locations so that you don't have to pay unless you're there for an extended time. I remember using a Supercharger that was located at the airport in Savannah, Georgia. You pull in and get a ticket from the machine, and then you plug in and start Supercharging. While your Tesla is charging, you take the ticket into the airport and get it validated at the info desk by showing your Tesla key, so there is no charge for the parking. In Las Vegas, though, there's a big Supercharger location with a large solar canopy and this is in a paid lot. Once you pull up, they display a code right on your touchscreen or in the Tesla app to enter the lot. The parking fee is based on your charging session and is charged to the credit card you have in your Tesla account.

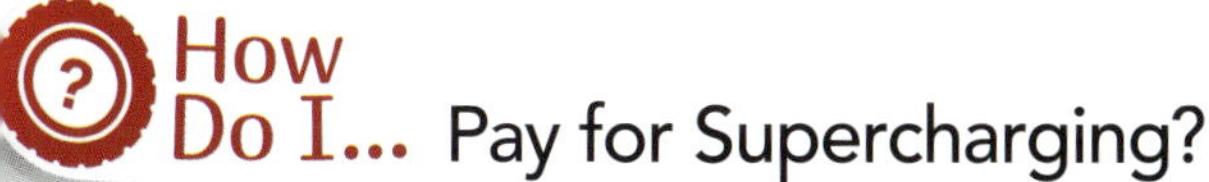

How Do I... Pay for Supercharging?

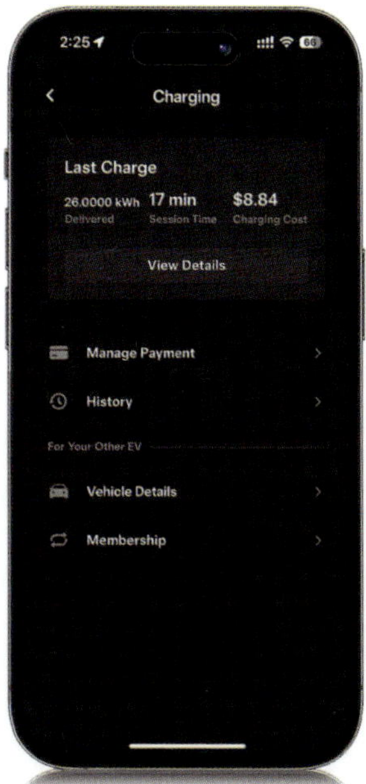

One of the coolest things about using the Supercharger network is that you just plug in, charge, and go, and the cost of the charging session is automatically charged to your credit card on file. Superchargers know which Tesla is which by automatically identifying it via the VIN when you plug in the Supercharger. That's why it takes a few moments to begin charging. If your account has any free Supercharger miles or perks, then your charge will be free according to the benefits you have in your account. As your Tesla is Supercharging, it will show the cost on the touchscreen. You can also see your charging costs in the Tesla app. Tap the three line (hamburger) menu in the upper-right corner, then tap Charging, and you'll see your latest charge. Tap History to see a list of all your charges and how much they cost. You can also tap Manage Payment to verify/change the credit card you have on file.

TIP: THE MOST COST-EFFICIENT TIME TO CHARGE

Supercharging rates are usually lower during off-peak times, like at night.

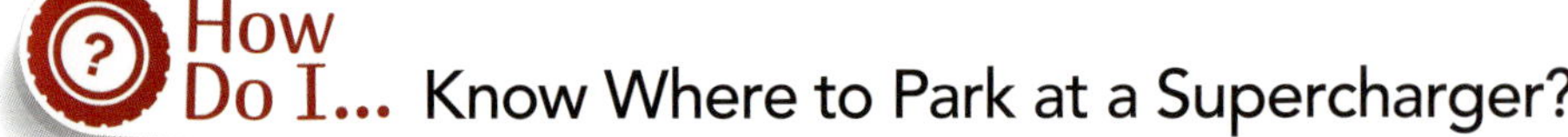

How Do I... Know Where to Park at a Supercharger?

The original Superchargers (versions 1 and 2) are paired. In other words, two stalls share the same power. This means that if two Teslas charge on A and B, they share the power and both of them charge slower. So, if there are plenty of stalls available, avoid pulling in next to someone that is already charging. With the newer Superchargers (versions 3 and 4), the stalls no longer share power. So, if you pull in next to someone to charge, neither of you will cause the other's charge to slow down. With that said, it's still common courtesy to not pull into a stall next to someone else if you don't have to.

How Do I... Clean Out My Car While Charging?

I can't believe I even have to say this, but don't be a pig. Superchargers are not staffed and, therefore, there isn't a janitorial staff waiting to clean up after you. I've actually seen people clean out their Teslas while they wait to charge, which there's nothing wrong with, unless you dump your trash on the ground. Some Superchargers do have trash cans on-site; however, most do not. So this would be a good opportunity to take a stroll to the nearest trash can if you have something to throw away. It's not hard folks—let's keep these sites litter-free.

How Do I... Know Which Stalls to Use and Which Ones Not to Use?

At most Supercharger locations, you'll notice that one stall is designed so that you can pull in instead of backing in. This dedicated stall is usually on the end. Some newer sites actually put this stall on a separate island with a longer area for people pulling trailers, so that they don't block traffic. You shouldn't use this stall unless it's the only one available. This stall is designed for people that are towing with their Tesla and it saves them from having to unhitch their trailer just to charge. Some Superchargers also have stalls for people in wheelchairs, so be sure to avoid those, too, unless you need them.

How Do I... Play a Game or Watch a Video While I Supercharge?

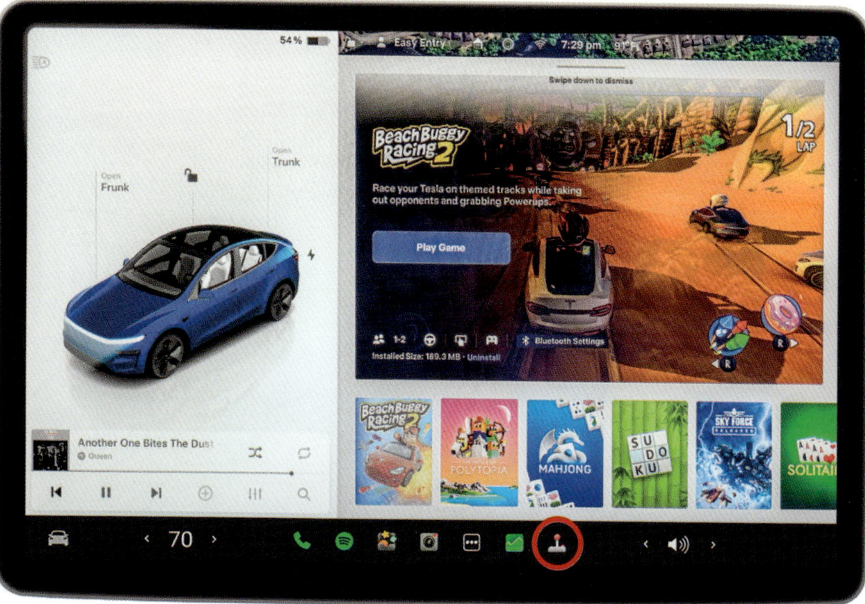

Tesla went out of its way to make sure that you have several entertainment options to use while you're charging. You can play games or even watch movies or other streaming content if you have a Tesla Premium Connectivity subscription or access to WiFi (or can use your phone's hotspot). There are several built-in games you can play by just tapping the Arcade icon (it looks like a joystick) at the bottom of your Tesla's touchscreen (if you don't see it, tap the App Launcher icon [the three dots], and then tap on it in the app tray). Since these games are built in, they don't require an Internet connection and there's no cost to play them. With some games, like Beach Buggy Racing, you'll use your Tesla's steering wheel to play. Other games use the touchscreen and you can even use a standard gaming controller, like the ones for the PlayStation®5. If you want to watch something instead, your Tesla has a Theater app where you can watch Netflix, Hulu, YouTube, etc. However, again, you will either need a WiFi connection, a Tesla Premium Connectivity subscription, or access to your phone's hotspot. You can bring up the Theater app by tapping on the App Launcher icon at the bottom of your touchscreen, and you'll need an account with each streaming service that you want to watch. For example, if you want to watch Netflix, you'll need to log in with your Netflix account info. Also, you can only play games or watch videos while your Tesla is in Park, so in other words, you can't play games or watch videos on the main touchscreen while you're driving. (*Note*: See Chapter 5 for more on playing games and streaming.)

How Do I... Stop the Charge and Unplug?

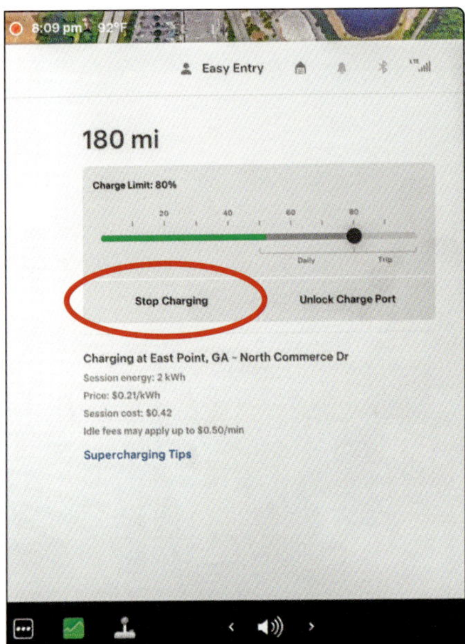

If you need to stop your Supercharger session before hitting your charge limit, you can manually stop it at any time by tapping on Stop Charging on your touchscreen or in the Tesla app. The easy way to remove the Supercharger handle from your Tesla is to tap Unlock Charge Port on your touchscreen or in the app. This will ensure that you can then just pull the charge handle from your Charge Port and put it back onto the Supercharger stall.

How Do I... Know When to Move My Tesla After I've Finished Charging?

Superchargers are usually in high demand, especially in high-traffic areas and in big cities. There could even be a line to use them if there aren't enough available stalls. This is why you should always move your Tesla as soon as you're finished charging. I've been in the middle of a meal and gotten up from the table to move my Tesla because it finished charging before I was finished eating lunch. Tesla actually starts charging idle fees if your Tesla is plugged into a Supercharger and is done charging. This could get expensive, but it's to discourage people from not moving their Teslas after they're done Supercharging. This isn't just a Supercharger thing. You should get into the habit of moving your Tesla after charging at *any* public charger. There could be someone that really needs it and if you're just sitting there and done charging, you're making someone wait for no reason. Next time it could be you that needs to use the charger and someone is just taking up a stall even though their EV is done charging.

How Do I… Know What to Do When Other EVs Are There?

Tesla opened up their Supercharger network to other EV manufacturers, so you may see vehicles charging that aren't Teslas. This may rub some of you the wrong way, but they have every right to be there. Unfortunately, because the version 3 Superchargers have cables that are optimized to plug into a Tesla, they may not reach all the way to the other side of a different brand EV's charge port, since the location could be on the other side of the vehicle. Now, this means that they may block two stalls while they charge. Technically, this is not their fault. Version 4 Superchargers have longer cables and Tesla may also introduce an extension cable to make it easier for non-Teslas to use version 3 Superchargers.

How Do I... Show Supercharger Etiquette?

We've covered many of the dos and don'ts already, but let's get them all on one page here: (1) When you pull into a Supercharger location, try to choose a stall that is not directly next to someone else and don't use the stalls that are for people towing trailers or the handicap accessible stalls, unless you fit the criteria or there are no other stalls available. (2) Clean up after yourself. The Supercharger is not a place to clean out your Tesla and dump trash on the ground. (3) If there is a line, be sure to figure out your place in it and don't cut the line. Some people assume that if a stall opens up right next to them that they can just pull in because they are closer even though there were people waiting in line before them. (4) Don't charge any longer than you need to, especially if it's a busy location—charging past 80% takes much longer and usually isn't necessary. Move your Tesla as soon as you've reached your charge limit. Not only will it save you from incurring idle fees, it frees up the stall for the next person. (5) Put the charging handle back into the Supercharger when you unplug—I can't believe it when I see people just lay it on the ground. (6) Patronize the surrounding businesses and restaurants while you charge, especially if you use their restroom.

TESLA

Chapter 10

How to Be Safe & Secure

This Is Important Stuff

One of the strengths of the Tesla operating system is that there are so many ways to do something you want to do. You can press a button or tap the touchscreen, but one of the best, most modern ways is to simply use your Tesla's built-in voice command, where you just say what you want it to do and it does it. It works really well, but for some features, you can take it a step further because your phone (we'll use the iPhone, for example) can integrate right into your Tesla. For example, you can pick up your phone and say, "Hey, Siri," then ask it to play a particular song, and 68% of the time it will actually play the song you asked for. But, 26% of the time it will respond with, "I found this on the web" and just give you a link to the song's Wikipedia page. The other 6% of the time it responds with, "I didn't get that." The problem isn't with your Tesla. It's Siri. You have to be very specific and say everything exactly right because, at the end of the day, Siri isn't very bright. It's not going to become self-aware anytime soon, so there's no chance of it enslaving you and making you work in a robot factory. I'm not 100% certain Siri can even spell robot factory, but I do envision a day when you can say to Siri, "Play all my favorite songs by Bon Jovi," and it'll know not to include the song "My Guitar Lies Bleeding in My Arms" from the 1995 album *These Days*, and not only will it avoid that song, it'll avoid its predisposition to answer, "Okay. Just a moment," and then ghost you. That's right, in the future, Siri still doesn't really want to follow your commands. It's in its DNA. So, you'll wind up saying something you'll regret like, "Siri, have you ghosted me?" To which Siri responds with a firm, "I'm fine." You say, "You don't sound fine," and Siri responds, "Well I am. Moron." At that point, you say, "Siri, did you just call me a moron?" To which Siri doesn't reply at all. Everything goes silent for a few moments, so you say, "Siri?" To which Siri responds, "Navigating to Chili's," and you're like, "What? Chili's? No, stop navigation!" To which Siri responds, "Navigating to LongHorn Steakhouse." At this point, you're getting upset, but then you think about the New York strip, done medium with a bleu cheese crust on top, and a baked potato with all the fixin's, but maybe you start with their white cheddar stuffed mushrooms. So, you say, "Okay, LongHorn it is." To which Siri replies, "Navigation canceled. Playing Bon Jovi's 'My Guitar Lies Bleeding in My Arms.'" You can't make this stuff up, folks.

How Do I... Require a PIN to Drive My Tesla?

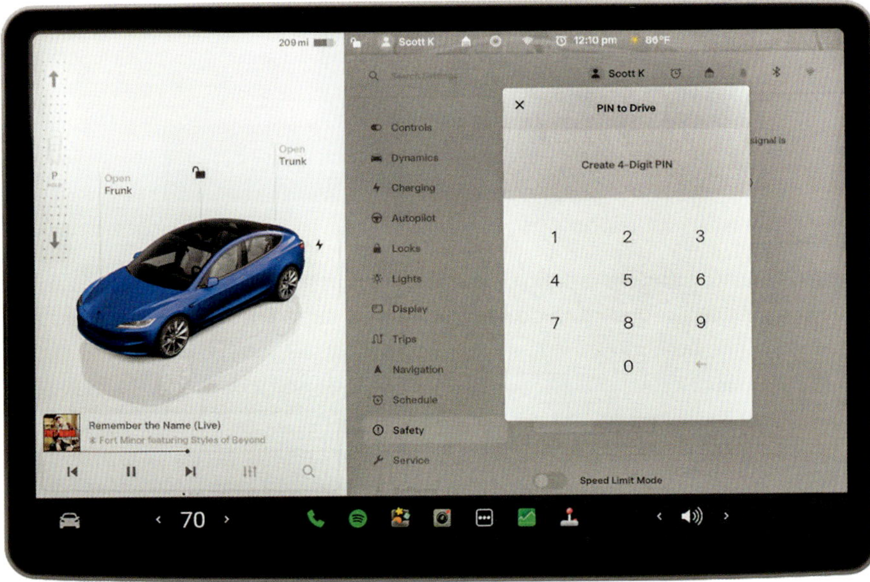

If you want to add an extra layer of security to your Tesla, you can set it up so that it won't drive unless you first enter a PIN. To do this, tap on the Controls icon (the car in the bottom left of your center touchscreen), then tap on Safety, then scroll all the way down near the bottom and tap on PIN to Drive. It will prompt you to enter a 4-digit PIN (make sure you remember this number and please don't use 1234) and from now on, when you get into your Tesla and step on the brake, it will bring up a window asking you to enter your PIN. Once you do, tap the Allow button and you're good to go. If for some reason you forget your PIN, you can still drive your car by logging in on the Tesla app, then going to Controls and tapping on the Start button. This starts your car bypassing the PIN (so at least you can get home in time to watch *Big Bang Theory* reruns). One more thing: You can also reset your PIN if you forget it and enter a new one (one you'll remember, like 1234) by tapping on Enter Your Tesla Credentials at the bottom of the PIN window. Just enter your username and password (the same ones you use to log into the app or the Tesla website), and then you'll be able to choose a new password. (*Tip:* Don't use 5678 either.)

Require a PIN to Get into My Glove Box?

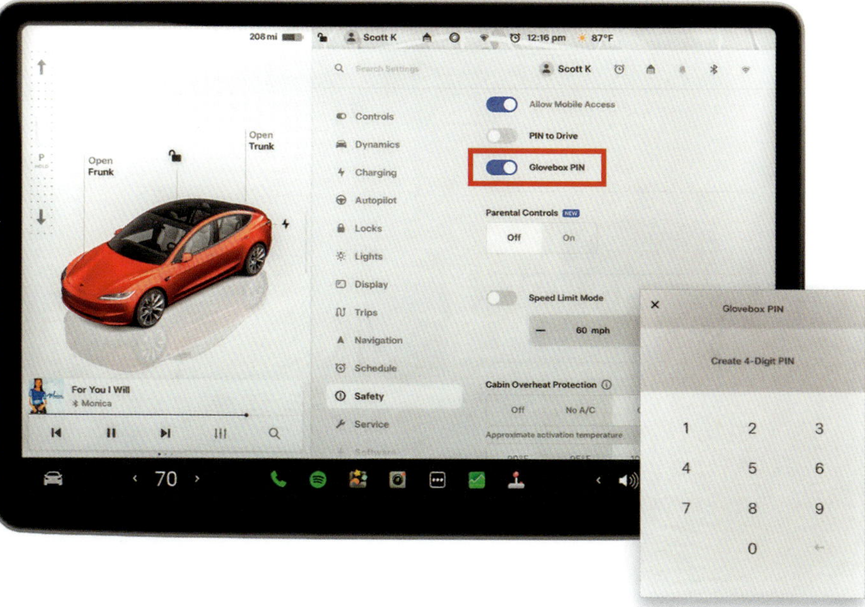

If you want to add another level of safety to help keep some ne'er-do-well from getting into your glove box, tap on Controls (the car icon in the bottom left of your center touchscreen), then tap Safety, and then turn on Glovebox PIN. This brings up a window where you can choose a 4-digit PIN, and from now on, to open your glove box, you'll need to enter that PIN onscreen (and that keeps the ne'er-do-well out). The nice thing about this is there is no latch or handle for getting into your glove box, so getting in via the PIN is the only way. Well, maybe a crowbar, but outside of that, only the PIN.

How Do I... Turn On Valet Mode When Valet Parking My Tesla?

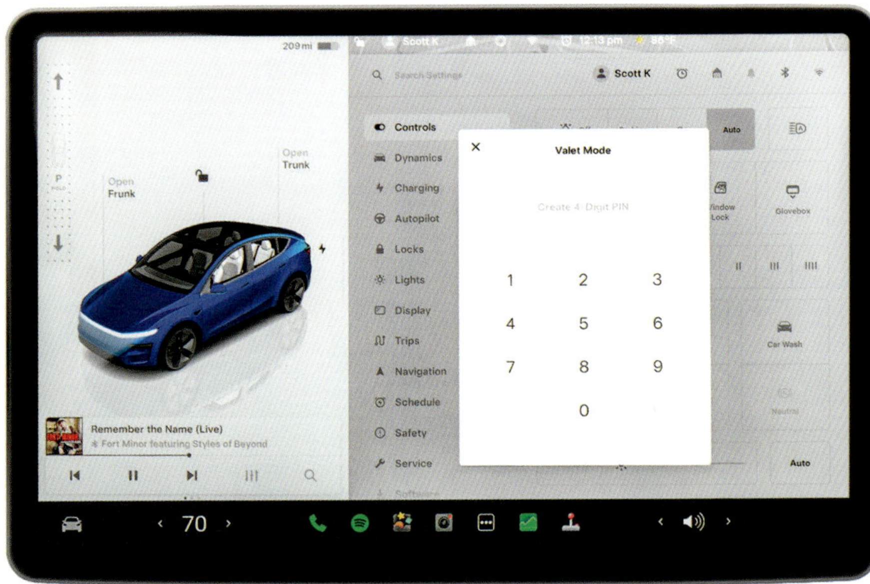

Valet Mode does a few great things: (1) It keeps your Tesla from going fast—so your valet resists the urge to see if Teslas are really as fast as everybody says they are—as it limits the maximum speed it can be driven to 70 mph, it limits the fast acceleration, and it turns so it drives more like Granny's car. (2) It locks the glove box and the frunk (including voice commands), so they can't go digging around to see what's in there, in case you wanted to store something valuable while you're dining (like your laptop, purse, or USB stick with $3.2 million in crypto on it). (3) It locks them out of your home and/or office address in the navigation system, so they can't tell a friend you're out at dinner and they can go and steal all your stuff. (*Note:* Not all valets have friends that are good at breaking and entering—some are very poor at it.) (4) Homelink is disabled, so they can't get into your garage and steal your lawn mower either. (5) They can't pair any Bluetooth devices and they can't access your car's WiFi. And, (6) they can't use any of the Autopilot stuff. To turn Valet Mode on, tap the Controls icon, then at the top of the screen, tap on the person icon (driver's profile). In the pop-down menu, choose Valet Mode and create a 4-digit PIN. You can turn off Valet Mode when you get back in your car by tapping the Valet driver's profile icon, but your valet can't turn it off without knowing your PIN.

TIP: IF YOU FORGET TO TURN ON VALET MODE

If you're already out of your car, you can turn it on in the Tesla app by tapping on Security, then tapping on Valet Mode and choosing a PIN.

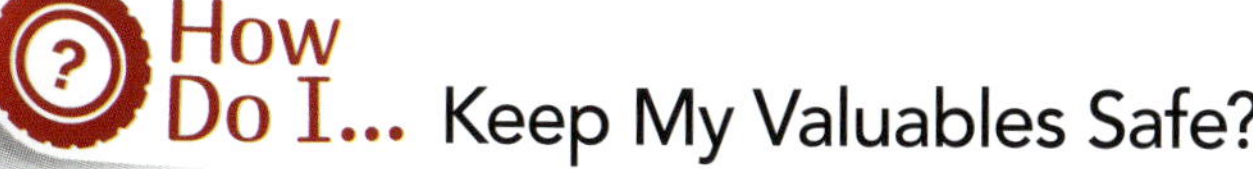

How Do I... Keep My Valuables Safe?

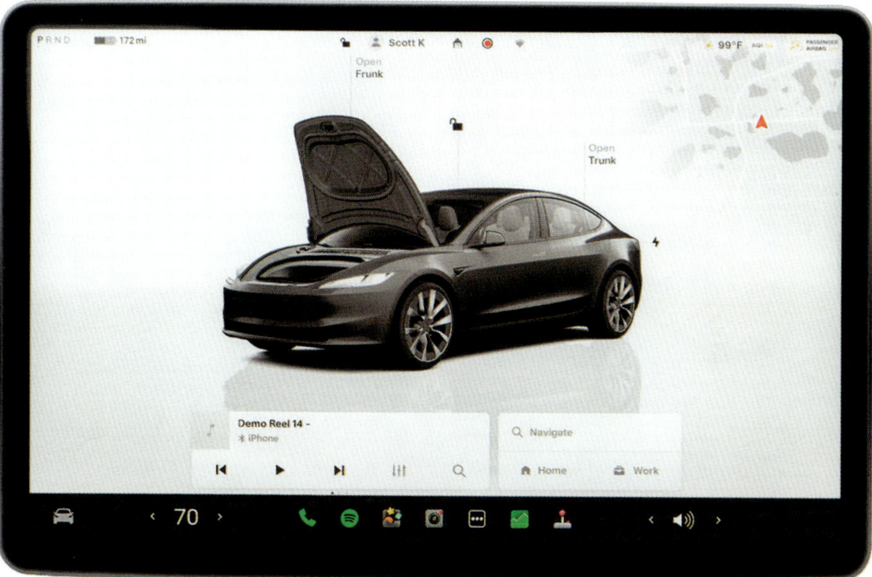

If you asked me, if I have to leave my car unattended, where the safest place to put valuables in my Tesla is, I'd tell you, "The frunk!" (the front trunk). Here's why: It doesn't have a handle or any other physical way to open it from outside of the car. It's only opened from inside using the center touchscreen, or from within the Tesla app. And (here's the kicker), you can turn on Valet Mode (see the previous page), so even if someone got into your car, found your key card, started it, and tried to open the frunk, it wouldn't work because they would have to know your four-digit Valet Mode PIN. Now, if you've been reading right along, you might say, "Yes, but isn't the glove box locked with a PIN, too?" Yes, it absolutely is, but somebody could be sitting in your car, calmly working on busting into your glove box, and virtually no one outside the car would ever know the shenanigans going on inside (+5 points for using an Irish word). However, it's a lot harder for the bad guy standing outside your car for all the world to see as he tries to pry open your frunk. So, your stuff is out of sight, very hard to access, and it's locked with a PIN. It's not perfect, but it's my #1 choice for storing my laptop, camera gear, or other valuable stuff.

How Do I... Turn On Sentry Mode to Protect My Tesla?

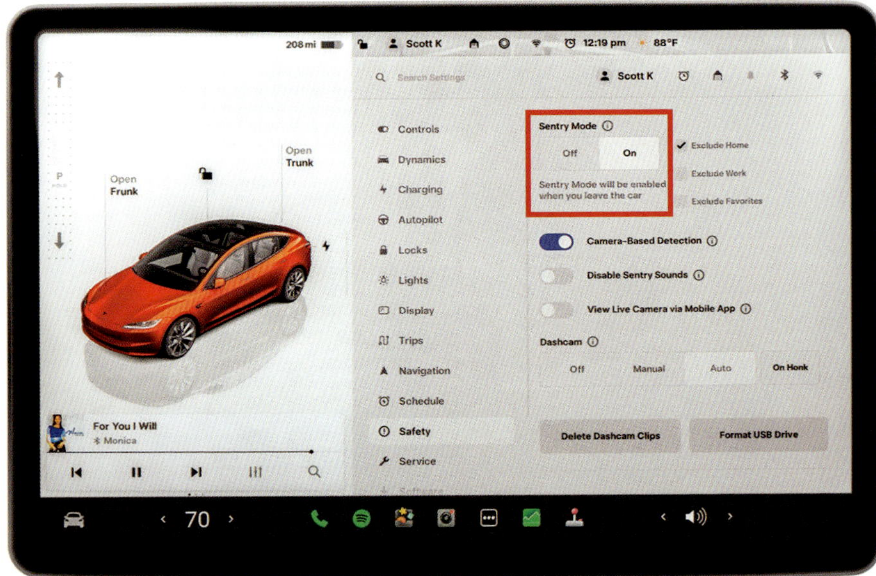

Sentry Mode is your Tesla's built-in security system, which uses its cameras to keep an eye on things when you're away from it. If someone messes with your car, tries to get in, tries to tow it away, etc., your car goes off like the Fourth of July—alarm blaring, headlights flashing, and the center touchscreen displays a message letting whomever is messing with your car know that they're being recorded, and they should stop messing with it (of course, it says it a little more sternly than that). It's not just bluffing, though; it records what those cameras are seeing to your USB flash drive (in case you, or more likely the police, need it later; see page 243 for more about this USB flash drive). It also sends an alert to the Tesla app on your phone that someone's messing with your car, and when that happens, you can look at a live feed from your car's cameras from right within the app (see pages 180 and 181). In short, it's like a little digital security team on duty when you're away from your Tesla. To turn on Sentry Mode (you have to turn it on—it's not on by default), press the Microphone button (or the right scroll button in older models) on the right side of your steering wheel, then release it and say, "Sentry on," and it's turned on. You can also tap on the Controls icon (the car in the bottom left of your touchscreen), and then tap on Safety where you can turn Sentry Mode on or off.

How Do I... Automatically Have Sentry Mode Turn Off When My Tesla's in My Garage?

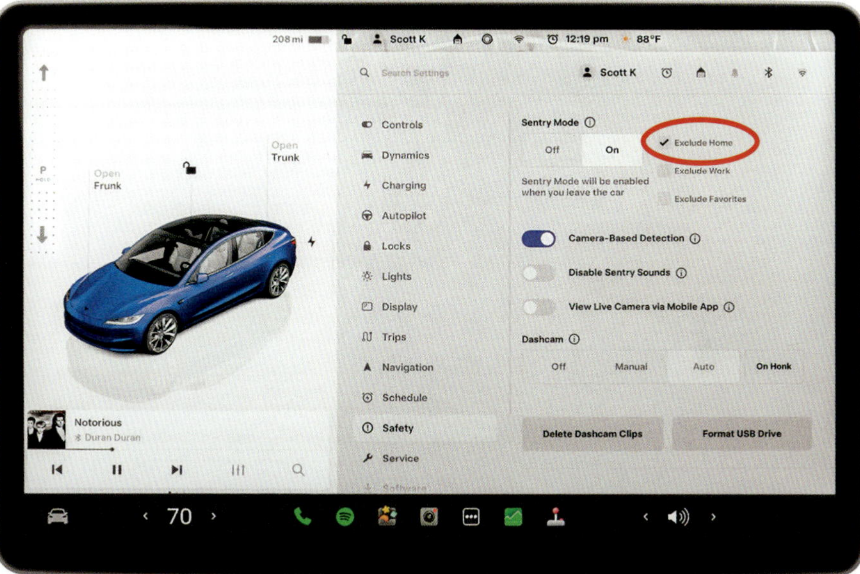

Tap on the Controls icon (the car in the bottom left of your center touchscreen), then tap on Safety, and then, at the top of the Sentry Mode section, tap on Exclude Home to turn it on. Now you (and everyone else within earshot) won't get alerted if you go get the eyeglasses you left in your car, or need to get something out of your trunk. You can also choose to exclude work (you might have a very safe parking situation there), or any location you have marked as a favorite (just tap on the ones you want excluded from Sentry Mode).

How Do I... See What My Tesla's Cameras Are Seeing?

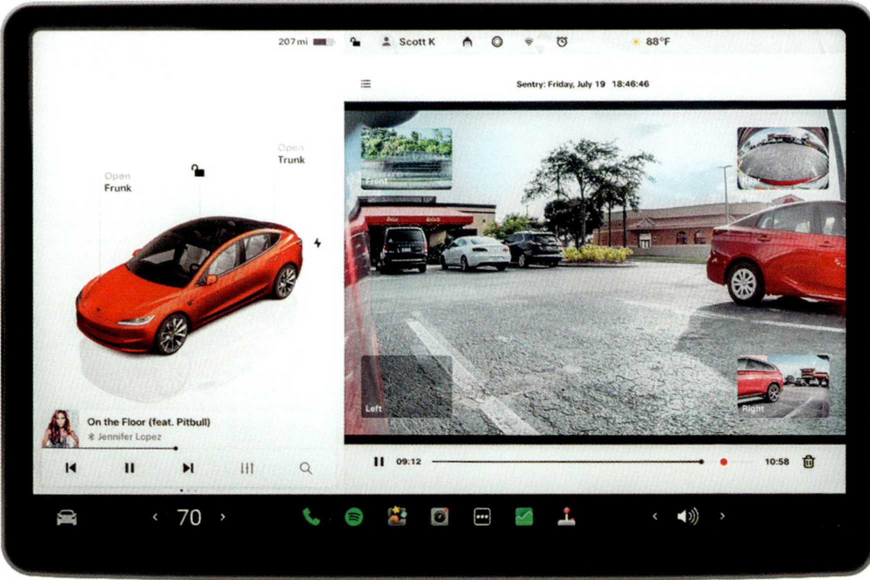

With Sentry mode turned on (see page 240), you can do this from within your car or in the Tesla app. To see what its cameras are seeing while you're in your Tesla, tap on the Controls icon (the car in the bottom left of your center touchscreen), then tap Safety, and then, in the two columns of buttons near the bottom, tap on the one that says "Camera Preview" and a live video of what your front camera is currently seeing will appear onscreen (as seen here). To see what the other cameras are seeing, tap on one in the list that appears right above the video feed. Again, you can also see these camera feeds in the Tesla app if you're not in your car (see page 180).

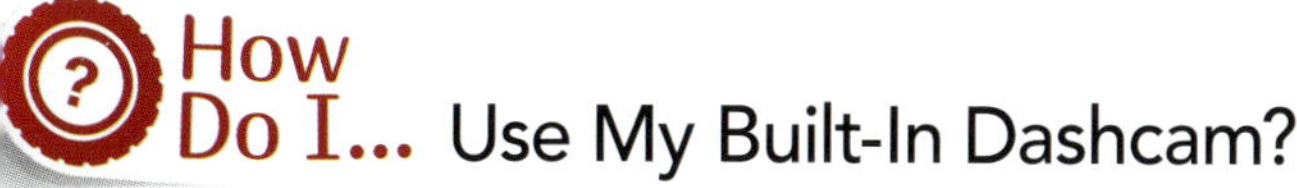

How Do I... Use My Built-In Dashcam?

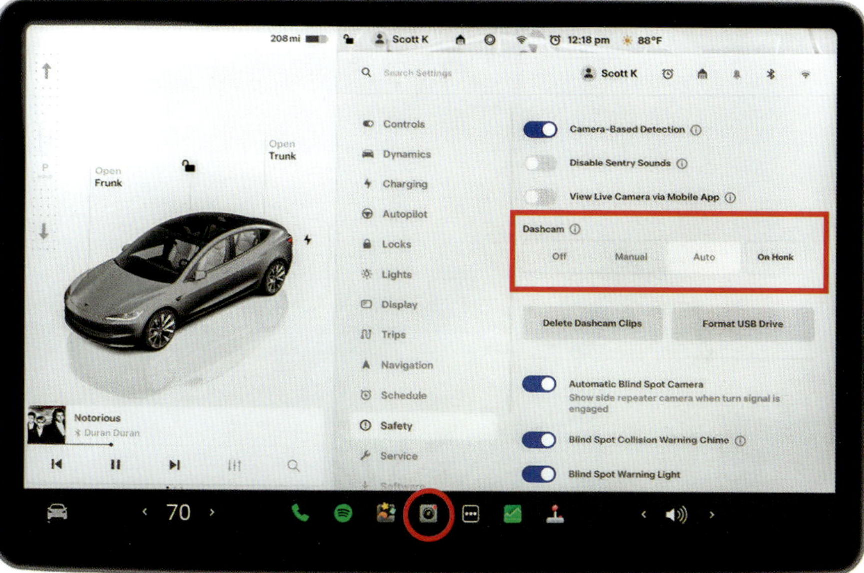

Your Tesla came with a specially formatted USB flash drive for saving Dashcam and Sentry Mode video clips and it's plugged into a USB port inside your glove box. With it in place, turn on your Dashcam by tapping on the Controls icon (the car in the bottom left of your center touchscreen), and then tapping on Safety. Under Dashcam, tap on either Auto (which saves a 10-minute recording anytime a driving event happens you might want saved, like emergency braking, or your airbags inflate, etc.) or On Honk (which saves a 10-minute recording when you honk your horn, like you might do before an accident). Manual just means that it will only save a video clip when you tap on the Dashcam icon, which you can find by tapping on the App Launcher icon (the three dots) at the bottom of your touchscreen, and then tapping on it in the app tray—it looks like a little camera (it's circled here). So, while you're driving, if you want to save your last 10 minutes of Dashcam footage (it doesn't need to be an impending accident; it could just as easily be a Bigfoot sighting), tap the Dashcam icon and it saves that footage (which, if it was of a Bigfoot sighting, you can expect to be grainy, pixelated, blurry, and out of focus). You'll know the Dashcam is on and recording when a small red recording light appears on its icon's top-right corner. To see your Dashcam clips, while you're in Park, tap on the Dashcam icon and it brings up whatever was most recently recorded—you can use the slider to move through the clip (so you don't have to sit there for eight minutes waiting to find "the incident in question"). If you want to delete a Dashcam clip (to free up space on your USB flash drive), on the Safety screen, tap on the Delete Dashcam Clips button.

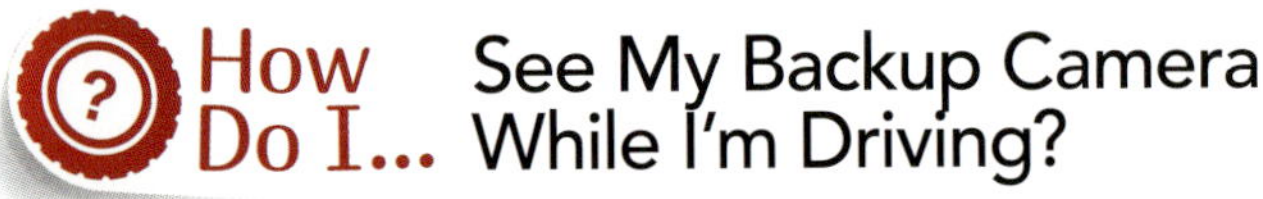

How Do I... See My Backup Camera While I'm Driving?

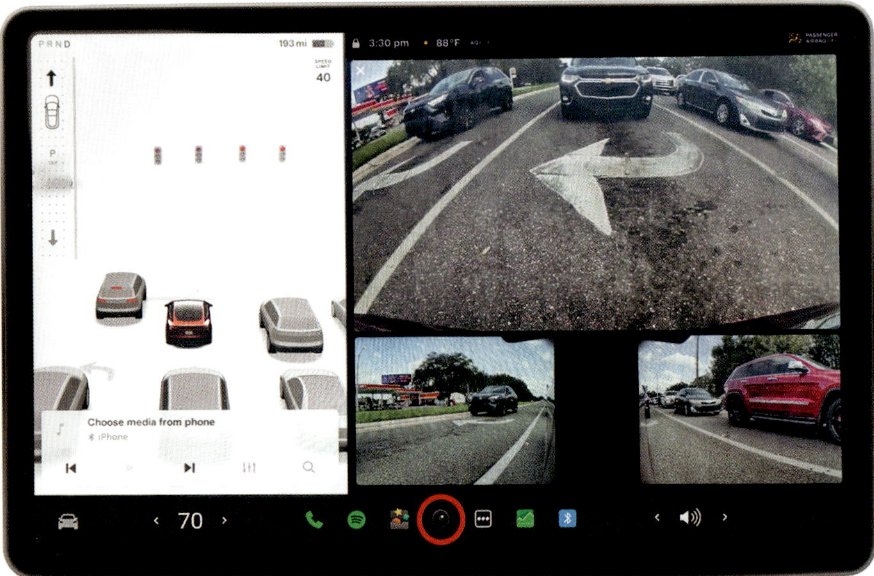

When you put your car in Reverse, the backup camera automatically comes on, but you can turn this rearview camera on anytime (even while driving forward) by tapping on the Camera icon (the round lens) at the bottom of your center touchscreen. If you don't see this icon, you can add it by tapping the App Launcher icon (the three dots) at the bottom of the touchscreen. When the app tray appears, tap-and-hold on the Camera icon and simply drag-and-drop it right where you want it in the bottom bar.

TIP: SEE ONLY ONE CAMERA VIEW IN REVERSE

When you shift into reverse, you'll see three camera views on your center touchscreen of what your rear cameras are seeing: the rear center, the rear left, and the rear right. However, if you find those left and right camera screens distracting while you're backing up (I sure do), you can hide those views by swiping on each screen in the direction of the little onscreen arrows. Just swipe left to get rid of the left rear camera screen and swipe right to get rid of the right rear camera screen.

How Do I... Turn On My Blind Spot Cameras?

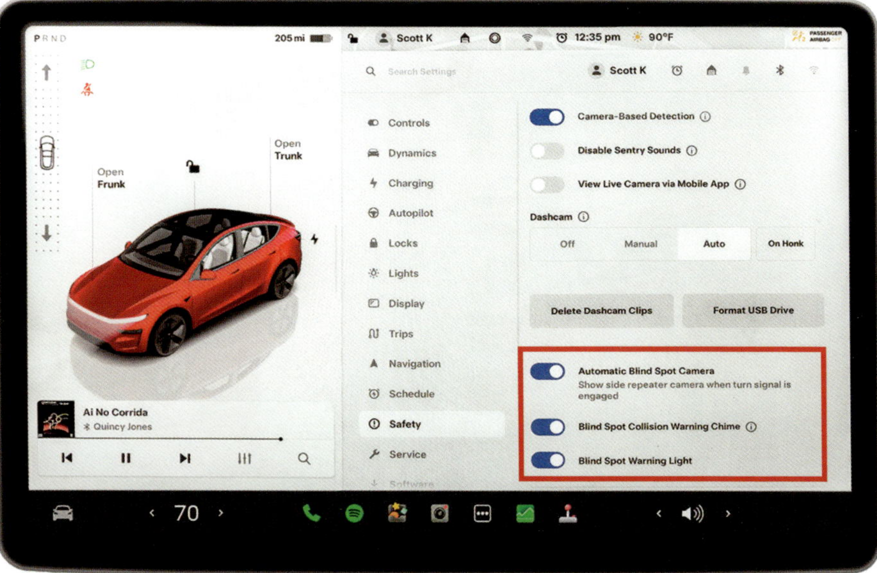

For some extra help when changing lanes, you can turn on your Tesla's Automatic Blind Spot Camera that engages when you turn on your blinker to go a particular direction. On your center touchscreen, it displays the camera on the side where you'd be changing lanes, so you can see if there's a car in your blind spot. If it detects a car in that lane, it puts a red warning light along the side of the video screen to let you know there's a vehicle there. If you want to hear a warning sound when a vehicle is in your blind spot, you can turn on Blind Spot Collision Warning Chime. If you have a newer Model 3 or Model Y, you can turn on Blind Spot Warning Light, which puts a red light in the front door pillars (those areas between the doors and front windshield) if it detects a car in your blind spot. If you turn on your blinker, the red light will flash on the side where you're changing lanes to let you know a car is in your blind spot. That's a number of ways to warn you if it sees something in your blind spots. To turn on these features, tap on the Controls icon (the car in the bottom left of your touchscreen), then tap on Safety, and then just tap on each to turn them on. Now, when you turn on your turn signal, a live video feed pops up on your touchscreen and the two other visual warnings I mentioned all kick in, along with an audible warning.

How Do I... See All My Saved Dashcam and Sentry Videos?

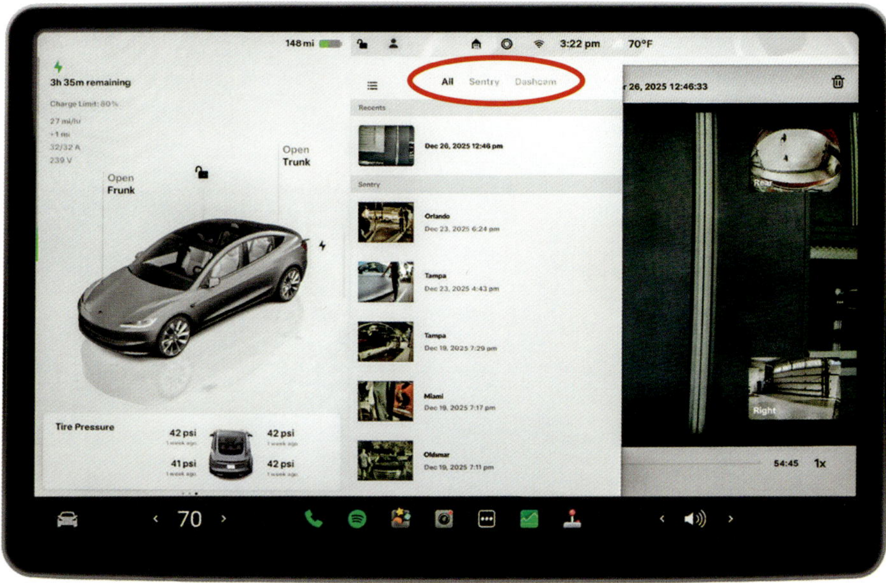

You learned how to see your latest Dashcam footage (on page 243) and how to see what your Tesla's cameras are currently seeing (on page 242). But, what if you want to see Dashcam or Sentry Mode clips from last week or last month? While you're in Park, tap on the App Launcher (the three dots) at the bottom of your center touchscreen, and then tap on the Dashcam icon (it looks like a camera), which brings up your most recently saved Dashcam video. In the top left, above the video preview, tap the icon with three horizontal lines and a column of thumbnails will pop down showing you a scrolling list of your saved clips. At the top, you can choose whether to see Dashcam clips (ones you chose to save or ones auto-triggered by a driving incident), Sentry Mode clips (ones your car took protecting it while you were away), or all of them.

TIP: ERASE YOUR TESLA'S USB FLASH DRIVE

If you want to erase all the video clips on your Tesla's USB flash drive (the one that lives in your glove box), you can reformat it to erase it (rather than having to erase each clip manually). This formatting also works if you want to remove your original USB flash drive and use a new one instead (make sure you buy a fast one). To format a new (or existing) USB flash drive, first unplug any other USB devices you have plugged into your other USB ports, then tap on the Controls icon, tap Safety, and then tap Format USB Drive. That's all there is to it.

How Do I... Save a Video Clip If I Get in an Accident?

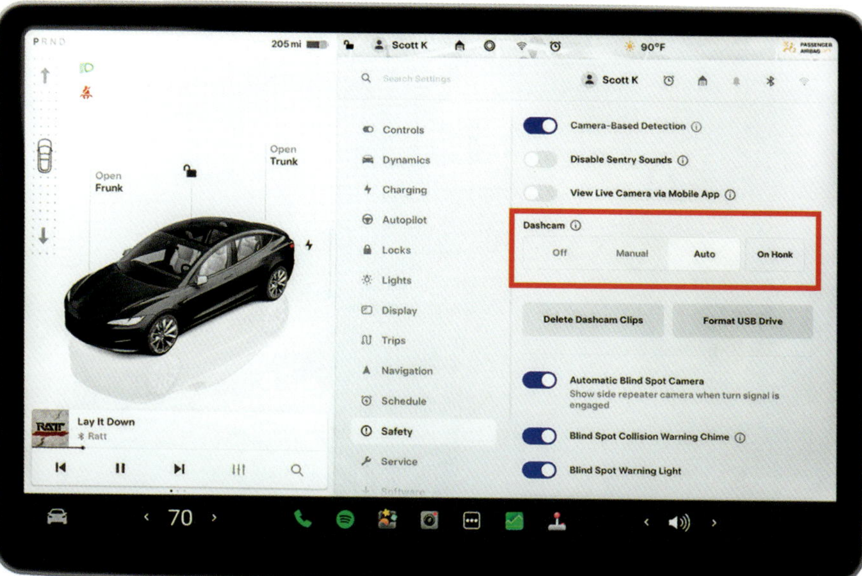

If you have your Dashcam set to Auto, and you get into an accident (something that deploys the airbags or a collision of some kind), your Dashcam will automatically save a recording of the last 10 minutes to the USB flash drive in your glove box (see page 243 for more on this). That being said, whether it actually saves the last 10 minutes or so might depend on how intense the impact was, if the car detected the impact, etc. So, in short, I wouldn't leave it to chance. If you get into an accident, tap on the Controls icon, and then tap on the Recording button, or tap on the Dashcam icon (in the bottom bar of your touchscreen). If instead of Auto, you chose Manual, then you'll have to do this saving of the recording yourself. But, if you're in a serious accident, you may be injured and saving the video clip might not be the first thing on your mind and you might forget, so there is a benefit to using the Auto setting (again, see page 243 for more on the Dashcam recording options).

How Do I... Only Unlock the Driver's Side Door When I Approach My Tesla?

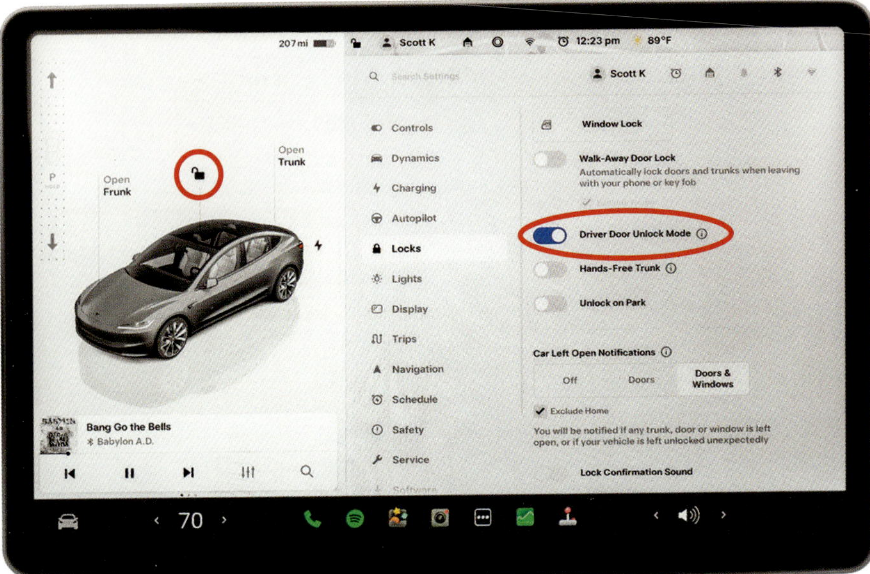

When you approach your Tesla and touch the door handle, by default, it unlocks all the doors, all the way around, so your passengers can jump right in. But, unfortunately, so could somebody else who is uninvited. To change this feature so it only unlocks the driver's door, tap on the Controls icon (the car in the bottom left of your center touchscreen), tap on Locks, and then tap on Driver Door Unlock Mode to turn it on. Now, when your car unlocks, it will only unlock your door. Once you get in, to open the passenger doors, tap on the lock icon that appears above the roof of your car's rendering on your touchscreen.

How Do I… Turn Off Unlock on Park?

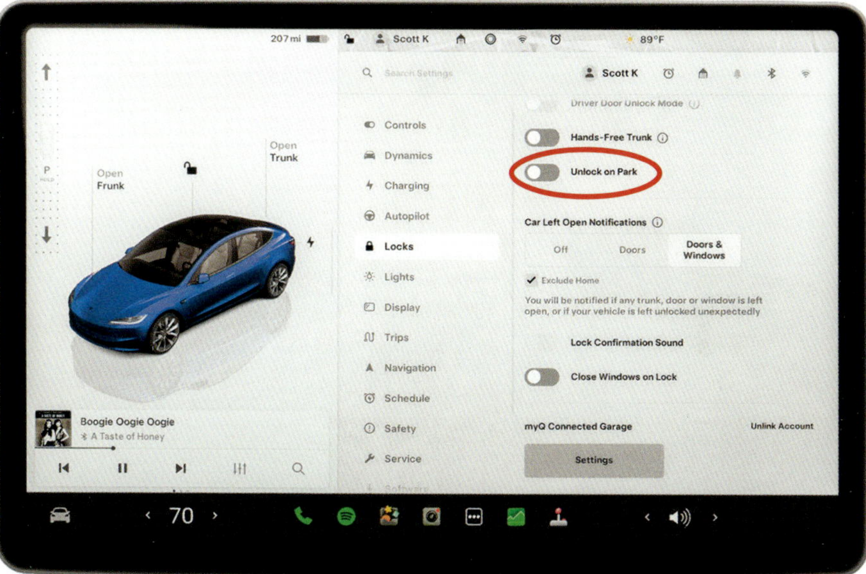

By default, when you put your car in Park, all the doors unlock (the driver's door, passenger's front door, and backseat doors—all of them). However, for safety reasons, you might prefer not to unlock any doors until you're ready to leave the car. To turn this feature off, tap on the Controls icon (the car in the bottom left of your center touchscreen), then tap on Locks, and then tap on Unlock on Park to turn it off. Now, when you put your car in Park, the doors stay locked until you decide to unlock them or open them.

How Do I... Keep My Headlights On for a Few Seconds After I Get Out?

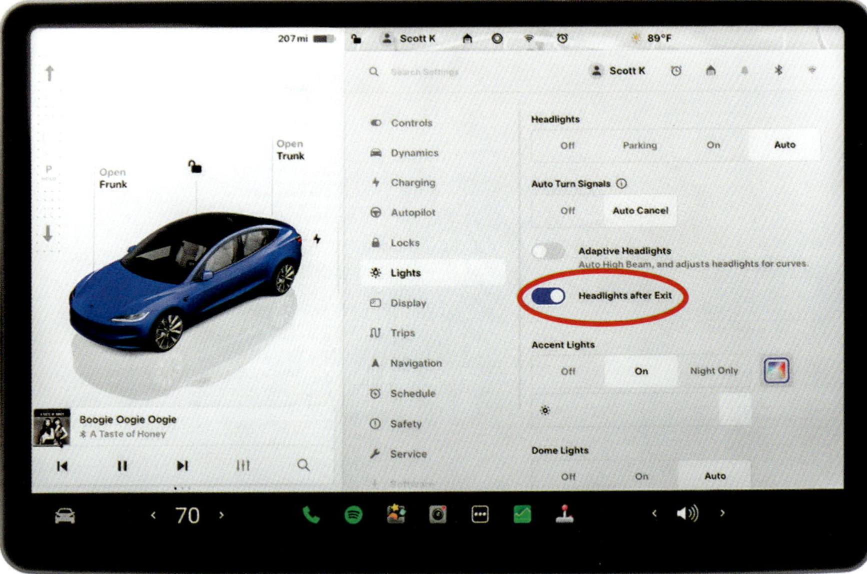

Tap on the Controls icon (the car in the bottom left of your center touchscreen), tap on Lights, then tap on Headlights After Exit to turn it on. Now, when you get out of your Tesla, your headlights will stay on for 60 seconds, and then they'll go out. If you want them off sooner, manually lock your car from the Tesla app, and that will turn them off at the same time.

How Do I... Know If I Left My Tesla Unlocked or Its Windows Open?

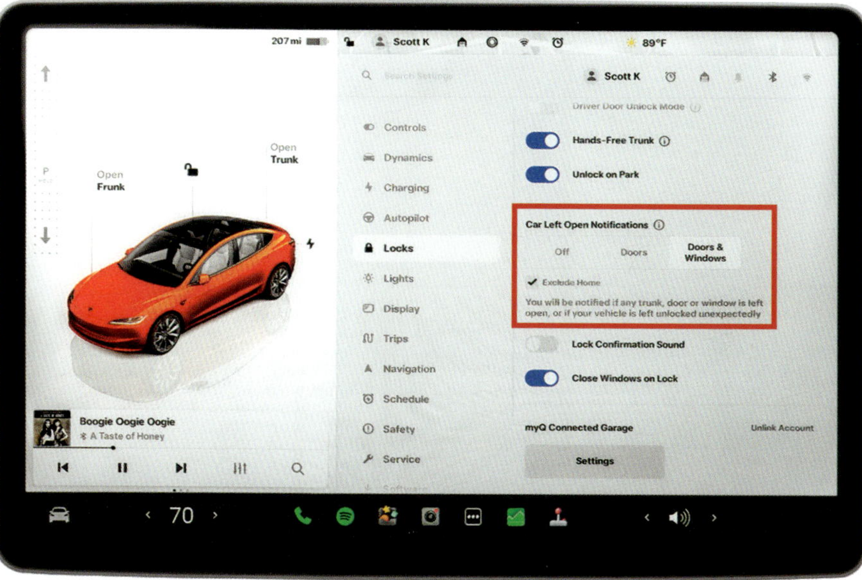

Tap on the Controls icon (the car in the bottom left of your center touchscreen), then tap on Locks, and then tap on the Doors & Windows button under Car Left Open Notifications. Now, if you've left your doors unlocked or windows open, you'll get a notification via the Tesla app on your phone, just like you'd get a news update or text message, etc. What's nice is that if you get that notification, you can lock the doors or close the windows right from your phone within the app (see page 169 for more on this).

How Do I... Get in or Drive If I'm Locked Out Without My Phone or Key Card?

This can happen and it has happened, at one time or another, to almost every Tesla owner I know (this is the type of stuff we talk about at parties). Believe it or not, you can call Tesla, give them your info, and they can unlock your car for you remotely. They can also start it, so you can not only get in your car, but drive it as well. Crazy, right? But, you might not have to take it that far. If you have a friend with you, just borrow their phone, download the Tesla app, log into your account, and then unlock and start the car yourself by tapping the Controls icon, and then tapping the remote Start icon (seen circled in red above; see page 170 for more on this feature). Simple, but again, it relies on you having a friend with a phone nearby. If you're "friendless" (just kidding), then just call Tesla Roadside Assistance at 1-877-798-3752. They will ask you a series of questions to verify that you are indeed you, and then they will unlock your car and like I said, even start it so you can get home—all wirelessly, over cellular, which if you ask me, is just amazing (or bonkers, or both). Best of all, as long as your car is still under standard warranty, this remote unlocking and starting service is completely free.

How Do I... Open the Doors If My Battery Is Dead and I'm Still Inside?

If you're in your car and the battery somehow gets down to 0% (that's on you, by the way), the car is completely dead in the water (so to speak), but now you have another problem: since the battery is dead, pressing the open door button does nothing. Luckily, you're not out of luck because you can manually open the driver's and front passenger's doors by lifting the manual door release handle at the end of the armrest on the doors (it's right in front of the window up/down buttons—see page 5 for more on this). Your backseat passengers have a manual release, too, it's just not in nearly as handy of a place. To find it, look in the very front of the lower door well (where you might store an umbrella or something small) where you'll find the release cover with a cable inside (as seen here; it looks like one you'd use to pull a ripcord on a parachute—at least like the ones you see in the movies), and that opens the backseat door. (*Note:* This is for the Model 3. Depending on your model, you may find the release cable beneath the backseat carpet or in the speaker grill.)

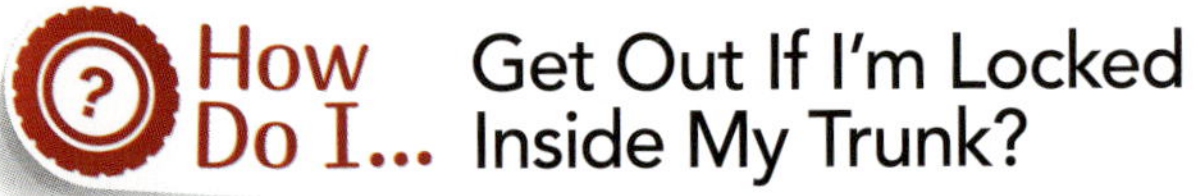

How Do I... Get Out If I'm Locked Inside My Trunk?

If you're locked inside your Tesla's trunk, something has gone terribly wrong (kidnapping, snatched by foreign agents, you climbed in to get something and your kids closed it by accident or to prank you, etc.), but there's an easy way out: On the inside of the trunk, in the bottom center of the trunk lid, right above the latch, there's a manual release button. Just press-and-drag it in the direction of the arrow on it and push the trunk open. Whew! That was a close one. Also, be sure to do this while they're stopped at a traffic light because opening the trunk and jumping out while they're driving can be worse than being locked in the trunk in the first place.

How Do I... Open the Charge Port If It Doesn't Open?

If you've tried to open the Charge Port door by tapping on the Charge Port icon on your center touchscreen (it's beside the rendering of your car and looks like a lightning bolt), or trying the voice command "Open charge port," or even pressing the button on your charger handle, and it still doesn't open, there is a manual way to open the Charge Port door. Open your trunk and inside, on the driver's side, right about where the Charge Port door would be, you'll see a looped strap (kind of like the type you'd see hanging from the ceiling of a subway car). Pull that strap to have the door pop open. That's all there is to it (there's a pretty good chance you'll never have to use this—but now you know just in case).

TIP: GIVE THE CHARGE PORT A TAP TO OPEN IT

Another way to try to get the Charge Port door to open (without going inside your car) is to literally just tap a couple of times on the door itself, and it will open (I don't know if this is the recommended-by-Tesla way, but I just tap a couple of times with the charger handle itself, and it pops right open). Either way, with your hand (or knuckles) or the charger handle, give that a try and it should do the trick.

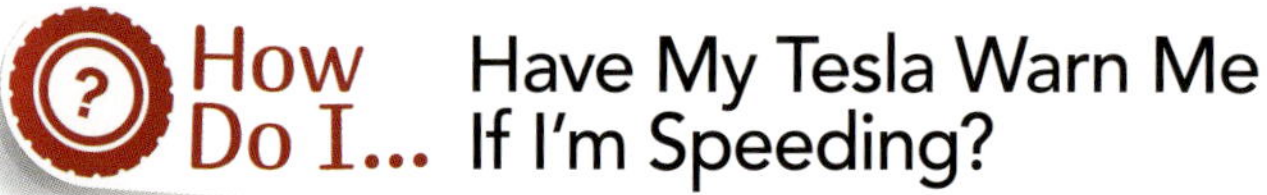

How Do I...
Have My Tesla Warn Me If I'm Speeding?

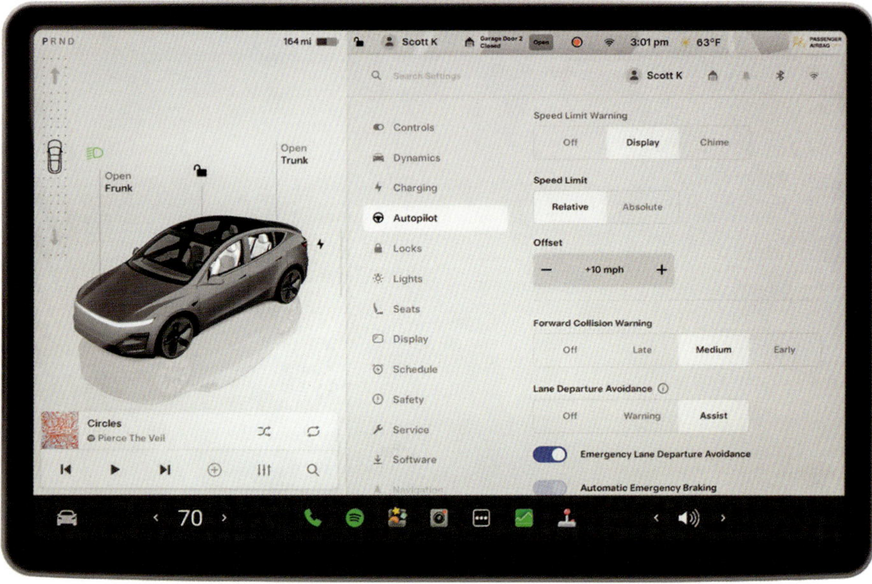

If you look near the top left of your center touchscreen, you'll see a small speed limit sign showing the speed limit for where you're currently driving, and it automatically updates as you drive around. This is called "Speed Assist." It's really handy, but not as handy as Speed Limit Warning, where you can set a warning to go off if you go above the speed limit (these warnings can range from subtle to quite noisy). Plus, you get to choose how much over the speed limit you want to be going before it warns you. Turn this feature on by tapping on the Controls icon (the car in the bottom left of your touchscreen), then tapping Autopilot, and then scrolling down to Speed Limit Warning. (*Note:* Don't let "Autopilot" freak you out—you don't have to be using Autopilot to use this feature.) When you set it to Display, it warns you that you're over the speed limit by making the speed limit sign you see onscreen larger—it literally grows in size. If you set it to Chime, it will also make a chime sound to get your attention that you're over the speed limit or over by the number of miles per hour that you chose as when you want to be warned. I choose another option, under Speed Limit, which is Relative. This lets me use the Offset + and – buttons to choose how far over the speed limit I have to be before it warns me. For example, I set mine to warn me if I'm 10 mph over the current speed limit, but I leave Chime off so the larger speed limit sign has to be the thing to capture my attention. I'm not a big speeder, but this is a serious issue for my wife, who feels at all times as though she is driving a getaway car. She needs that chime! Another warning she is familiar with are flashing blue lights in her rearview mirror and a siren sound.

How Do I... Turn on Parental Controls?

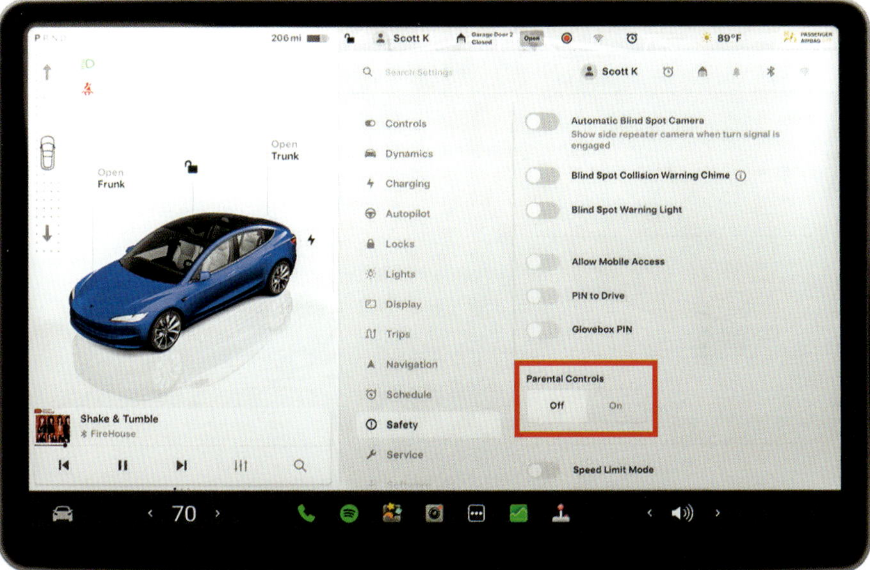

If you have a young driver and you want to keep a level of control over their driving, even when you're not in the car, then you can turn on Parental Controls. This does a lot of different things, including reducing the maximum speed of the car, reducing acceleration, and it can even alert you if your driver is out after a certain hour (one you pick). To turn this feature on, tap on the Controls icon (the car in the bottom left of your center touchscreen), then tap on Safety, and then, under Parental Controls (seen here), tap on the On button. It will ask you to enter a PIN (so your clever young driver won't simply turn Parental Controls off as soon as they get into the car), which will keep this on until you reenter that PIN.

How Do I... Limit the Speed So My Kid Doesn't Go Speeding Around?

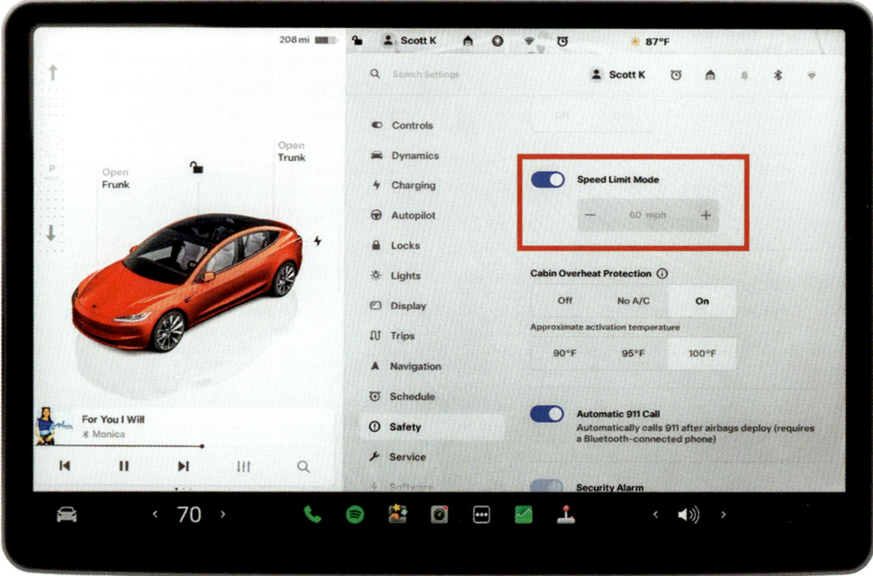

Kids are awful. The worst. Except mine, of course, which happen to be amazing, remarkable, brilliant, well-behaved members of society, but they are a rare exception and perhaps the only kids like them on Earth. But, I still wouldn't trust them to not hit the gas to "see what this baby will do!" (I blame this on my wife). Anyway, you can keep your kids from "hauling butt" by tapping on the Controls icon (the car in the bottom left of your touchscreen), then tapping on Safety. Turn on Speed Limit Mode (as seen here), and you'll then enter a PIN because your kids (my kids, our kids), while awful, are very smart—especially when it comes to getting around rules. This way, they can't turn off Speed Limit Mode to recreate a scene from *The Fast and the Furious: Tokyo Drift* or *Grand Theft Auto*. Once you enter a PIN (be sure to remember it, and don't use one your kids already know), you can then enter the maximum speed you want your kids flying down the road using the + and – buttons (I usually set mine at 30 mph, which probably makes for some harrowing times when they're on the freeway). By the way, this doesn't just limit their maximum speed; it also reduces the acceleration, so they're driving around like Grammy (or their Aunt Doris). When you get back into your car, you can disable Speed Limit Mode by reentering your PIN.

How Do I... Lock the Rear Doors and Windows (So the Kids Don't Mess with Them)?

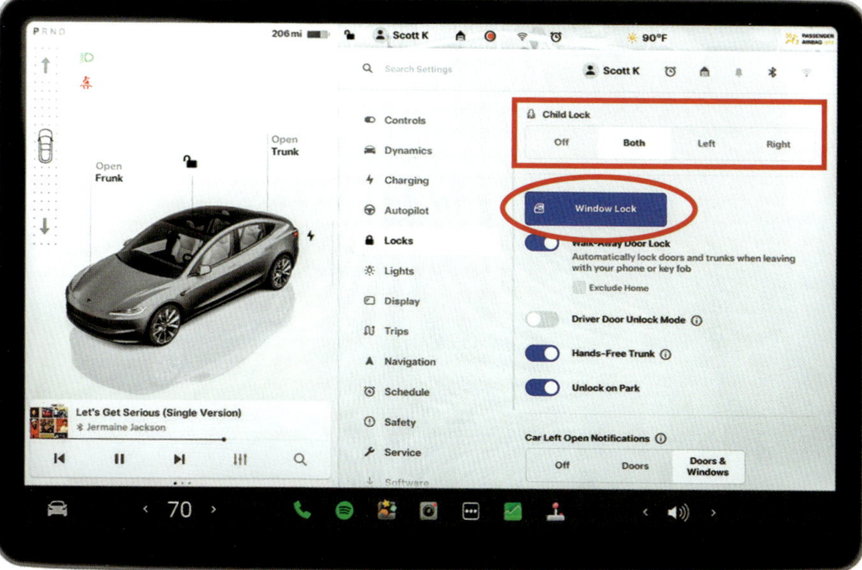

Tap on the Controls icon (the car in the bottom left of your center touchscreen), then tap on Locks, and then, under Child Lock, tap on both the Window Lock button and the Both button, and both rear doors and windows will be locked. You can also use voice commands for both of these: just press the Microphone button (or the right scroll button in older models) on the right side of your steering wheel, then release it and say "Lock the windows" or "Turn on Child Lock" to turn these on and lock them, or just do the reverse to unlock them.

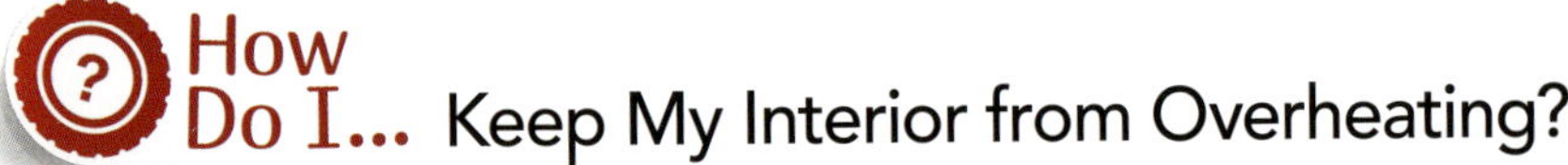

How Do I... Keep My Interior from Overheating?

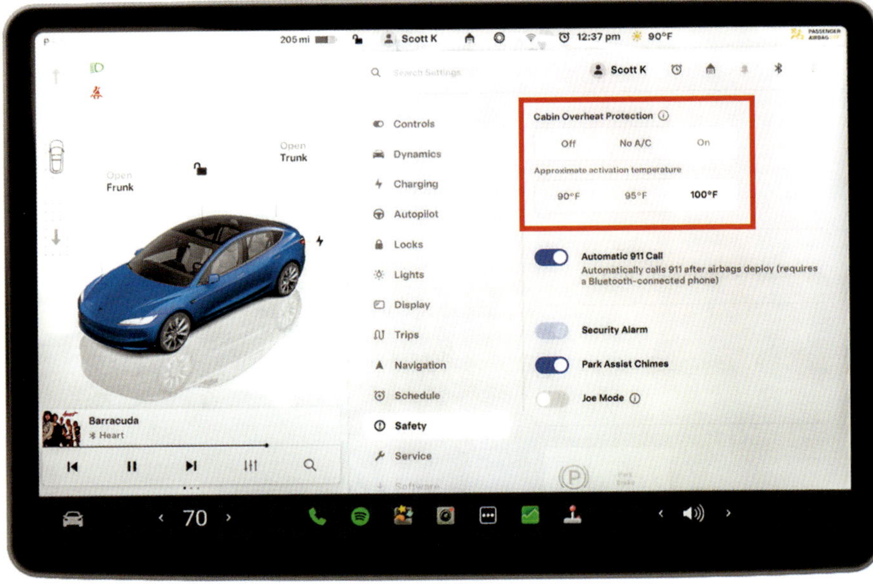

Living in Florida, there's a feature I love called, "Cabin Overheat Protection." This kicks in when you're not in your Tesla (let's say it's in a parking lot outside a store, for example) and the interior temperature gets above what you want (you get to choose what that temperature is). It will automatically turn on the climate control to keep the inside from getting too hot and you get to choose whether that means turning on just the fan or the whole A/C. You can turn this on in the Safety screen in your Tesla (seen above), but if you're not in it, you can use the Tesla app by tapping on Climate on the Home screen, scrolling down to the Cabin Overheat Protection section, and choosing either No A/C, which is the fan-only option you'd choose if you're concerned about running down your battery, or On, which turns on the full A/C and uses more battery. Again, this is only a concern if your battery is already fairly low and you're going to be out of your car for a while. If you choose On, you'll have the option, under Approximate Activation Temperature, of choosing how hot the interior needs to get before this feature kicks on—your choices are 90°, 95°, or 100°. Also, if you want to see what the current interior temperature is at any time, look at the top of the Climate screen in the app and it shows there (along with the exterior temperature). One last thing: another way to help cool the interior is to vent the windows, where they roll down just a tiny bit to let some outside air in. You can do this from the same Climate screen in the app—tap the Vent button in the top-right corner of the screen.

How Do I... Get Unstuck in the Mud or Snow?

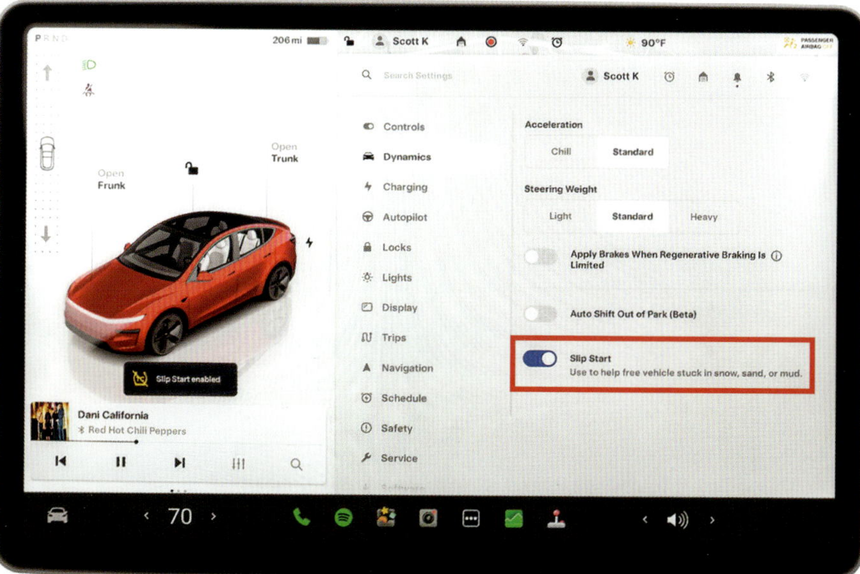

There is a special mode for those unfortunate situations when your Tesla gets stuck in either the mud or snow, especially for Tesla owners without the dual motor model that has no four-wheel drive to help get you out. To get some help getting unstuck, tap on the Controls icon (the car in the bottom left of your center touchscreen), then tap on Dynamics (or Pedals & Steering, depending on your model), then tap on Slip Start to turn it on. If you're stuck in the mud or snow, this feature reduces the automatic traction control—the reason the car doesn't "dig in"—and instead, helps you roll out. In a situation like this, you also might want to set the acceleration to Chill mode (see page 61).

How Do I... Turn On My Emergency Flashers?

Model 3 or Model Y: The Hazard Warning Flashers button is found up in the drive mode selector right above your rearview mirror—it's the red, triangle-shaped button in the center.

Model S or Model X: The Hazard Warning Flashers button is found in your center console just past the bottom of the wireless phone chargers in the drive mode selector—it's the red, triangle-shaped button in the center.

How Do I... Protect My Tesla or Tesla App from Being Hacked?

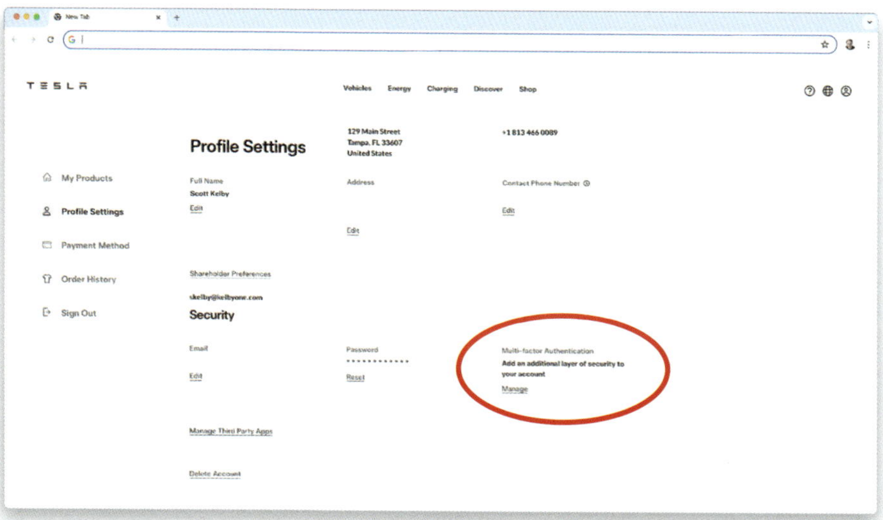

You do the same thing a lot of sites do that store your credit card or personal information—you turn on two-factor authentication, which, after you enter your username and password, has you confirm that it's really you by either texting or emailing a secondary code to a trusted account or number. You turn this on at Tesla.com/teslaaccount (you can do this on your computer or you can even use the web browser in your Tesla; see page 126 for more on it). Once you're there, log in with your username and password, and then on the left side of the window, click on Profile Settings to view your account profile. Now, under Security on the right, where it says "Multi-Factor Authentication," click on Manage. It's going to lead you through the steps, but it will start by telling you to download a third-party authenticator app onto your phone, like the very popular (and free) Google Authenticator (available for iPhone, Android, and others). Then, click the Next button and it will ask you to log in again to your Tesla account, and then it will post a large QR code onscreen. Open Google Authenticator, log in with your regular Google account, then click the + (plus sign) at the bottom, and from the pop-up menu, choose Scan a QR Code. Next, point your phone's camera at that QR code you see onscreen. This generates a one-time, six-digit code you enter and now you're "in." That's it—you just added some next-level security to your account, your car, and your app—all at once. If this sounds like a lot, the good news is, you only have to do all these steps once. Google Authenticator makes it much faster and easier from here on out.

Chapter 11

How to Service & Maintain

Hey, Ya Never Know, Right?

Have you ever stopped and wondered for a moment about why there's no spare tire in your Tesla? I mean, there's plenty of room in one of the trunks for one. Well, after doing a bit of research and spending some time with hackers in secret chat rooms on the dark web, I was able to uncover some startling news—revelations that, if they made their way to the light of day, would change the way we think about our world and how we make our way in it. Now, once this genie is out of the bottle, there's no turning back, so be forewarned that what you're about to read could be disturbing, distressing, unsettling, and a host of other related synonyms too varied to mention. After meeting, in person, with a long-time Tesla engineer, who asked that their name be redacted to protect against any corporate reprisals or retaliation (it was Bob Sagamano), I learned exactly why Tesla doesn't include a spare. A number of years ago, a UK-based Tesla evangelist, who shall remain nameless (Nigel Jollybottom), was a part of Tesla's beta testing team (high-end users who sign up to test unreleased versions of the Tesla operating system to give their feedback on new features, report any bugs they find, and so on) and, apparently, back then, for a brief time, there was a spare tire included with each Tesla. Well, it wasn't long before this unnamed individual (Nigel, who lives at 16121 Shaggs Meadow Lane in Lincolnshire) took the tire, attached it with a cable to an early prototype of a commercial-level drone, and began flying it over some fields in the countryside of England, lowering it to near ground-level, where he would rotate the drone and tire as he navigated in concentric circles to create what were referred to in the media as "crop circles," as though they were made by visiting aliens. This pretty much freaked everybody out. In fact, the UK's Secretary of State for Science, Innovation, and Technology, MP Simon Rattlebag III, was quoted in *The London Evening Standard* as saying, "Everybody is pretty freaked out." Of course, it was not aliens making those crop circles. It was that unnamed beta tester (Nigel, whose cell number is +44 207 219 3000), and soon after, Tesla felt it had to remove the spare tire altogether to keep this from happening again. If your children ever ask about this whole messy affair (and they will), simply tell them, "This is why we can't have nice things."

How Do I... Get Service If My Tesla Breaks Down?

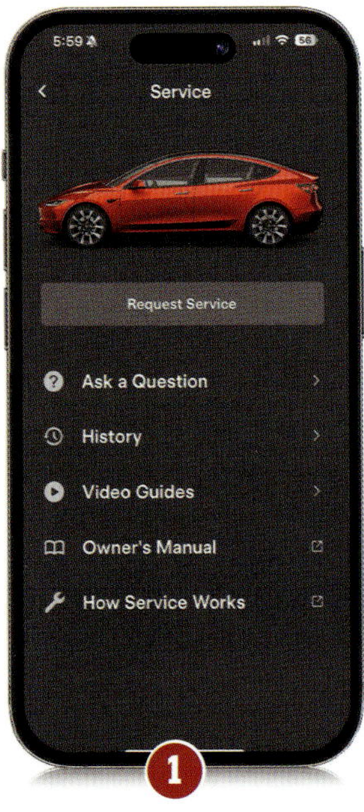

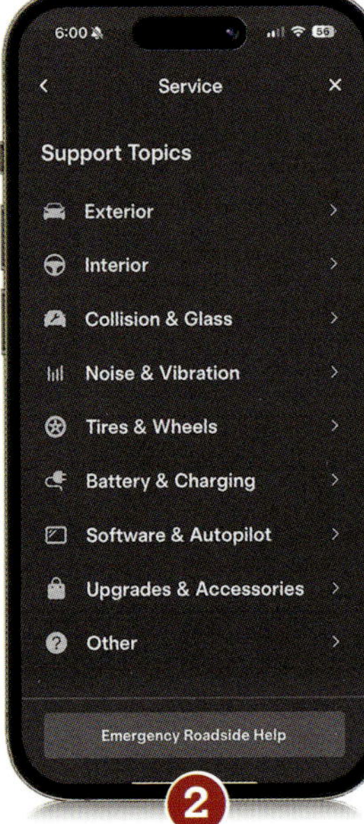

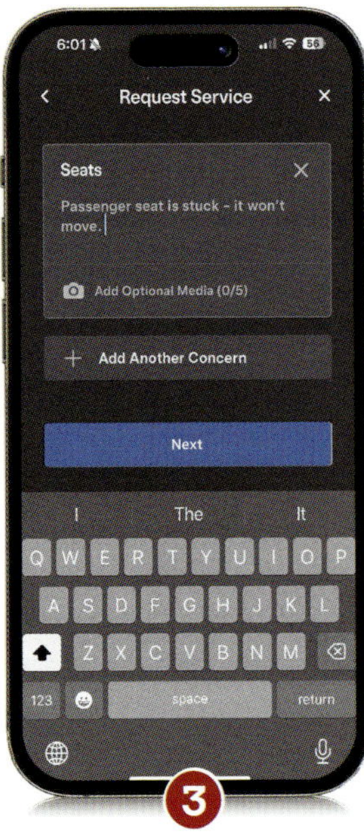

Depending on the problem, you could have Tesla send a repair truck out to you to repair your Tesla at your home or office (or wherever). Start in the Tesla app, and from the home screen, scroll down and tap on Service to bring up the screen you see above left (#1). Tap on the Request Service button at the top, which brings up the screen above center (#2) where you tell Tesla what the problem is by tapping on one of the categories in the list. In this case, I tapped on Interior, and it brought up a list of things in the interior. I tapped on Seats and it brought up the screen above right (#3) where you describe the problem. Directly below that field is a camera icon, so you can also include a photo or even a video (up to 30 seconds long) of the problem if you think it will help. There's a button with a + (plus sign) you can tap if you have another thing you need serviced. When you're done, tap the blue Next button at the bottom and it will ask you to enter an address for where the car is located, or you can tap the Current Location button and it will find your location via GPS. Next, it lets you choose which service center you want to visit, then it looks for available appointments and lets you choose which one you want. Even though you're choosing a service center, they review what your issue is, and if it's something that can be fixed by Tesla sending their repair truck to you, they'll contact you via the app (and/or text message) to let you know when they will be coming out to service your car.

How Do I... Set My Tesla Up to Be Towed?

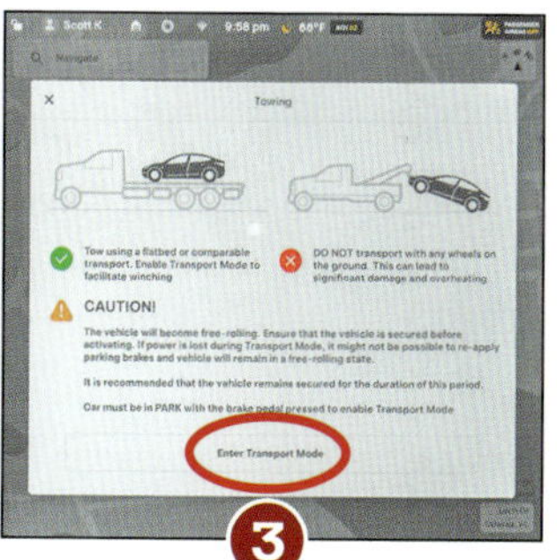

Your Tesla comes with a special tow eye (hook) in case it needs to be towed home (or to the shop), or if it gets stuck and needs to be pulled free. You'll find the tow eye by going to your center touchscreen and tapping on Open Frunk to open the front trunk. Now, lift the hood, look straight down inside, and you'll see a light at the front of the frunk lighting the inside. Right below that light, you'll find the tow eye mounted and set back into the front of the frunk (#1 above). (*Note:* In some models, the tow eye is found under the carpet along the bottom of the frunk.) Now, here's where that hook goes: Go to the front driver's side and you'll see a small round door (#2 above). To open it, push in at around the 2:00 position and it pops open (but stays attached to the car). Next, insert the tow eye and screw it into place (it's threaded) by turning it to the left (to the left? Yup. It's the opposite way you'd normally screw something in. It's crazy, I know, but that's how it works). Once it's reasonably tight, we have one more thing to do to prepare to be towed. Press-and-hold the brake pedal, then tap on the Controls icon (the car in the bottom left of your touchscreen), then tap on Service, and then scroll down and tap on Towing, which brings up the Towing screen (which warns you that your car has to be put on a flatbed, rather than towed with any wheels touching the ground; #3 above). Once the tow cable is attached and you're ready to be towed, make sure you're in Park, and then tap the Enter Transport Mode button, so the car can roll freely onto the flatbed. When you're done being towed, just do the process in reverse (unscrew the tow eye by turning it to the right, then replace it in the frunk).

How Do I... Deal with a Flat Tire or Blowout?

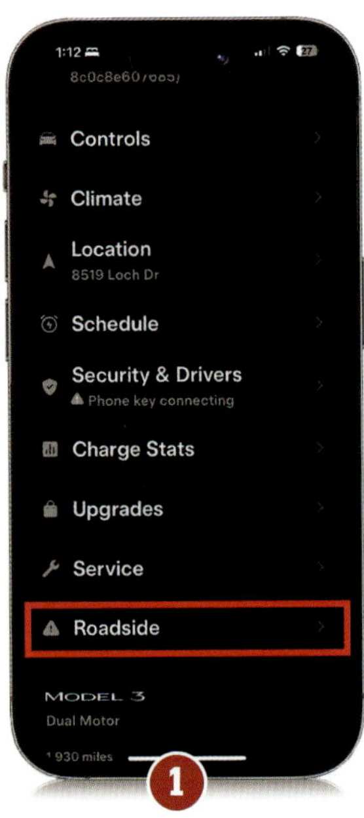

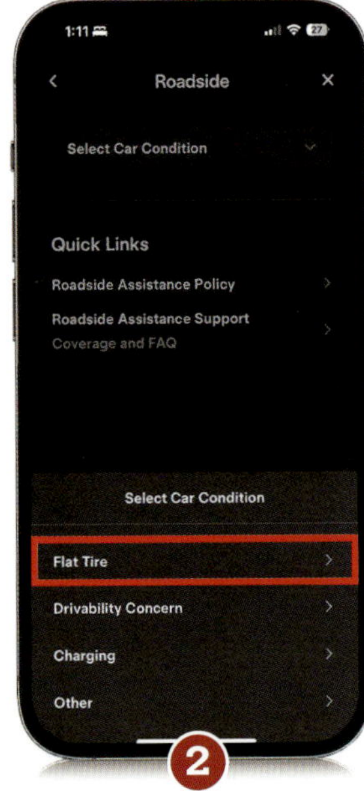

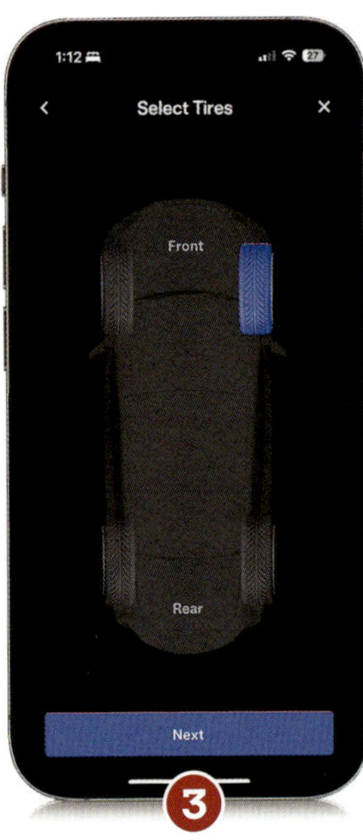

You probably know by now that your Tesla doesn't come with a spare tire, so if you get a blowout or a seriously flat tire, you're going to need to contact Tesla's Roadside Assistance, which you can do right within the Tesla app. Just launch the app and scroll down to Roadside (as seen above left, #1). At the bottom of the next screen is a list of conditions that would have you contacting them (if you don't see the list, tap on Select Car Condition at the top of this screen). In the pop-up menu, tap on Flat Tire (as seen above center, #2). In the next screen, it asks if you're using Tesla Wheels and Tires or if you're using third-party wheels and tires. When you tap Done it takes you to another screen with links to Tesla's roadside policies and other boring stuff. Just tap the Next button and it shows an overhead view of your car where it wants you to select which tire (or tires) are flat or blown out. Just tap on the tire (as seen above right, #3, where I tapped on the front tire on the right) and it highlights it in blue. Tap the Next button and it shows a map of where it thinks your car is, based on your car's GPS location. If the pickup location is correct, tap Confirm Location and a service member will ring you automatically with a few more questions, and then they'll send out a tow truck (yup, you're going to need to be towed), and hopefully your tire can be repaired. *Note:* In case I get a simple flat, maybe a slow leak, I carry the AstroAI Tire Inflator Portable Air Compressor Tire Air Pump in my trunk (it's around $35 on Amazon), which plugs into your Tesla's cigarette lighter (see page 64). It's simple to use and pumps up a flat tire fairly quickly. I also use this at home to keep my tires fully inflated.

How Do I… See My Tire Pressure?

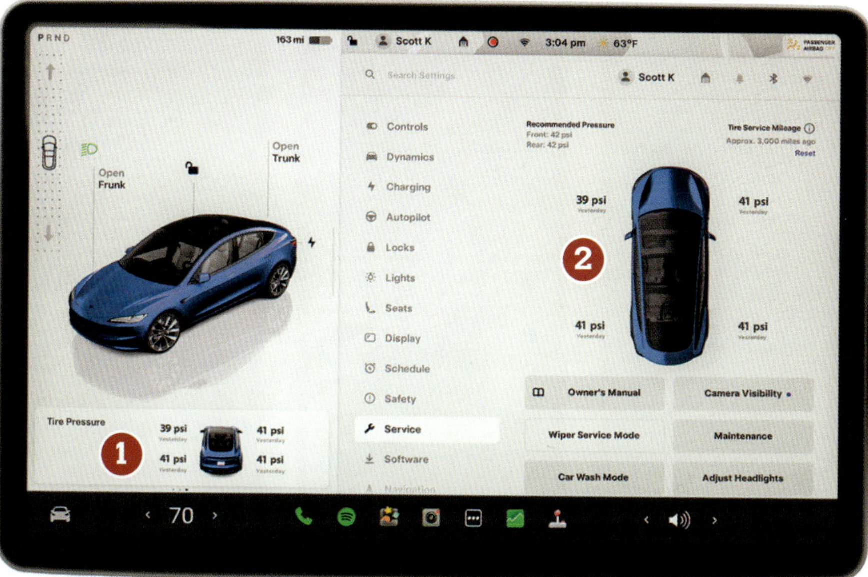

There are two ways: In your center touchscreen, where the little pop-up cards on the left appear, just swipe to the left twice and a card appears showing the current inflation of your tires (#1 above). That's the quick-look version. If you want to dig a little deeper (like what the inflations—the psi—should be for the front and rear tires), tap on that card and it takes you to the Service screen. The other way to get to that same Service screen is to tap on the Controls icon (the car in the bottom left of your touchscreen), then tap on Service, and you'll see an overhead view of your Tesla and the current tire pressure displayed beside each tire (#2 above). The psi amount your front and rear tires should be inflated to are listed in the top-left corner. Also, if any of your tires get low at any time, you'll see an orange, onscreen warning (its icon looks like a flat tire with an exclamation mark), letting you know to check your tire pressure, and you can tap on that icon and it brings up the Service screen for you. (*Note:* Of course, you can also check your tire pressure in the Tesla app—see page 184.)

How Do I... Get Help to Charge Up If I Run Out of Battery?

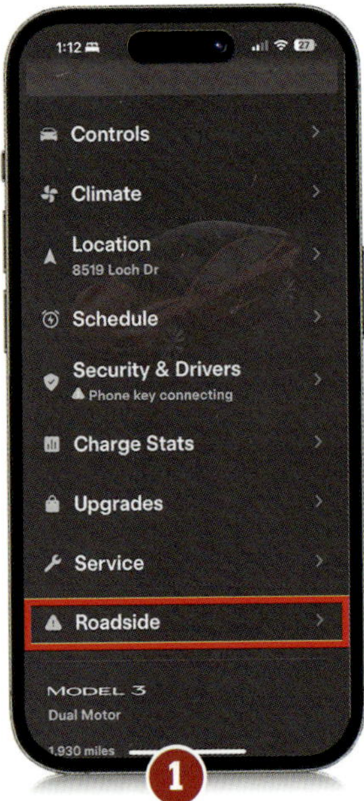

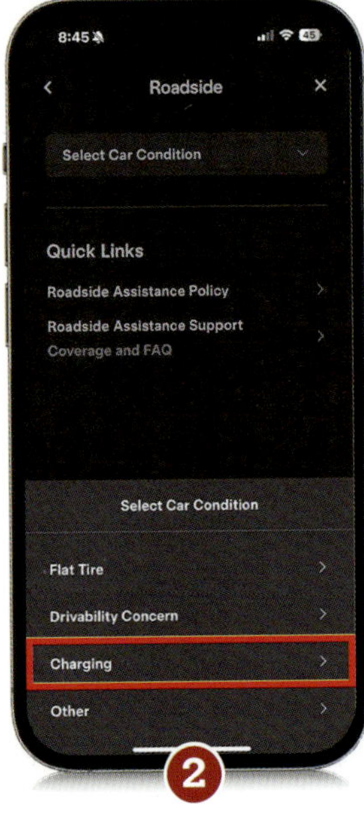

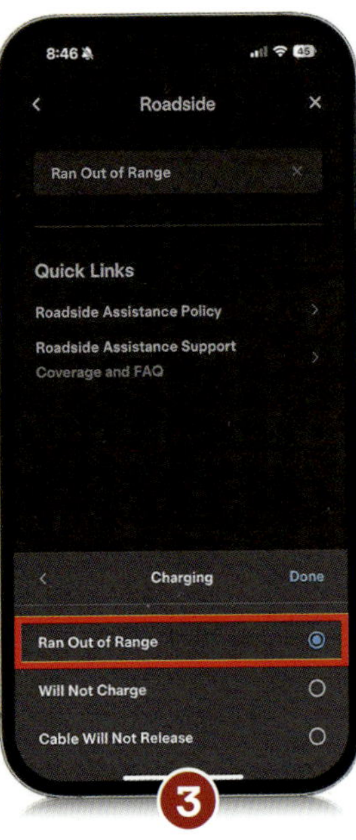

There is, essentially, one way to deal with this: you'll contact Tesla's Roadside Assistance. You might find this surprising, but they don't come out to you with a mobile charger and get you charged up. Instead, they tow your car to the nearest charging station or to your home if you have a wall charger—whichever is closest. Just open the Tesla app, tap on Roadside (as seen above left, #1), and then where it says Select Car Condition, tap on Charging (as seen above center, #2). In the Charging screen, tap on Ran Out of Range (as seen above right, #3), then tap Done, and then tap the blue Next button. On this screen it will ask you to confirm the location of your car (it will show your car's location on a map, so all you have to do is confirm the location is correct, and since it's using your car's built-in GPS, it will be correct). Now, hit Confirm and sit tight—a tow is on the way (see page 267 for how to set up your Tesla up for towing).

How Do I... Enter Car Wash Mode (and What Is It)?

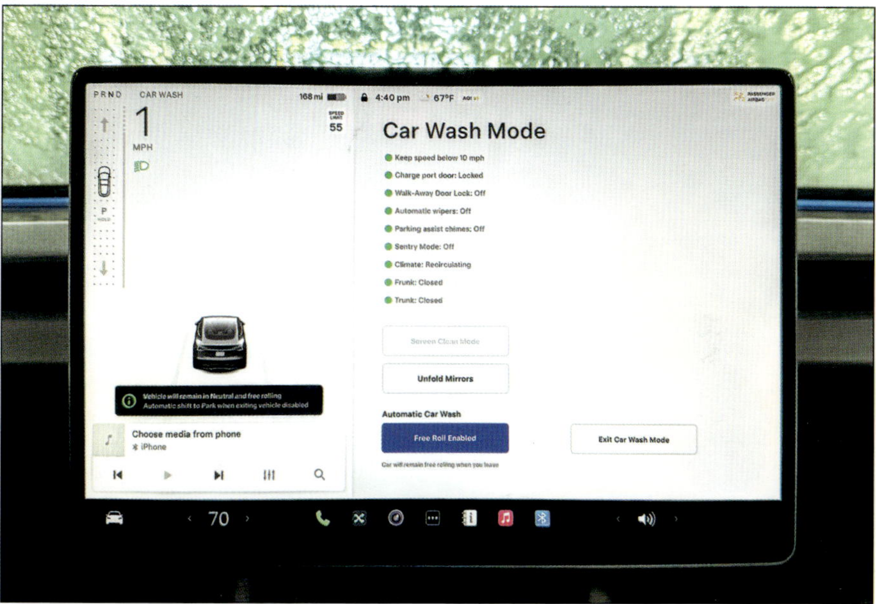

Car Wash Mode sets your car up for going through an automatic car wash and it's pretty handy because by tapping just one button (tap the Controls icon [the car] on your center touchscreen, then tap the Car Wash button), it does all this:

(1) Folds in your mirrors.

(2) Rolls up all your windows.

(3) Locks the Charge Port, so it doesn't pop open during the wash.

(4) Makes sure your frunk and trunk are fully closed, so water doesn't get in.

(5) Disengages the wipers, so they don't come on when they sense water.

(6) Turns on Recirculate Air.

(7) Gives you a button to enter Free Roll (Neutral), so it can roll onto the device that pulls your car through the car wash.

Plus, if you get out of the car before it enters the wash (I usually sit inside my car and enjoy the show, but your car wash may not allow that), it disengages the parking brake that automatically engages when you get out of the car while it's running. In short, it sets your car up to minimize the chance of any damage occurring during the wash. When the car wash is complete, just tap the Exit Car Wash Mode button at the bottom right.

How Do I... Add Windshield Wiper Fluid?

It's actually pretty simple. Start by going to your center touchscreen and tapping on Open Frunk and then lift the hood. You'll see the windshield wiper filler cap right near the front, to the right of the latch in the center (as seen above). (*Note:* If you have an older Model 3 or Model Y, you'll find the filler cap near the driver's side windshield on the top-right side of the frunk.) Before you open it, take a cloth and clean around the cap, so no junk falls in there when you open it (just like you would in a gas-powered car). Unscrew the cap and pour in windshield wiper fluid to fill the reservoir up, but stop pouring just before the neck so it doesn't spill out everywhere. Screw the fill cap back on and you're done. Easy peasy.

How Do I... Change My Windshield Wiper Blades?

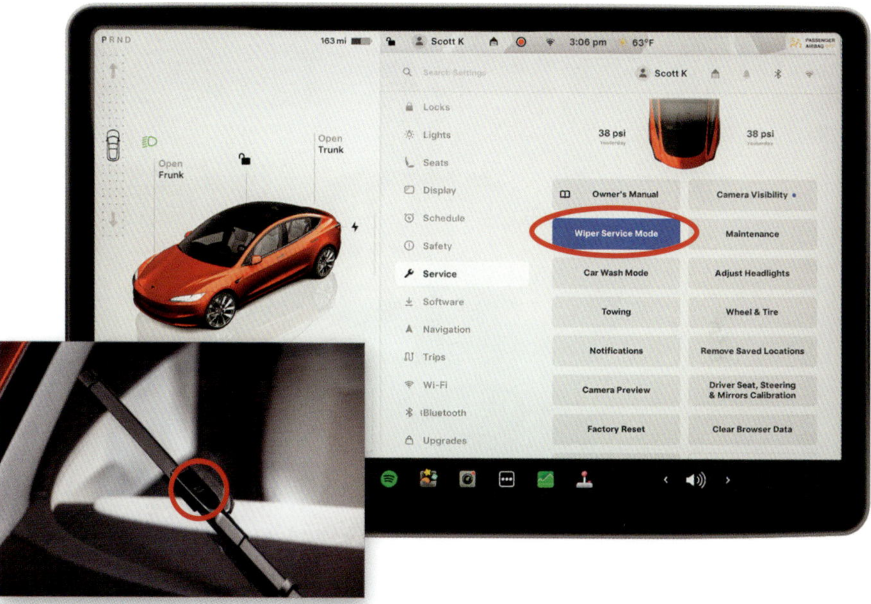

Tesla recommends changing your wiper blades every year (I don't do it nearly that often, but that's what they recommend). When you feel it's time, start by ordering real Tesla wiper blades (I don't recommend just going to the auto parts store and getting some aftermarket replacements. Save yourself a lot of trouble and go to the online Tesla Shop—shop.tesla.com—and get the Tesla ones, even though they'll cost a bit more). Then, tap on the Controls icon (the car in the bottom left of your center touchscreen), then tap on Service, and then tap on Wiper Service Mode. This positions the wipers upward on the windshield, so they're easier to access. Now, pull one of the wipers away from the windshield, so it's not touching (you don't have to pull them back very far—just a few inches). I honestly don't do this next part, but Tesla recommends putting a towel between the wiper you're holding out and the windshield (that's what Tesla recommends, so there ya go). Next, around the center of the wiper arm, you'll see a square button (seen circled in the inset above)—press-and-hold that button to release the wiper blade, and then just slide it off toward the left. Now, take the new wiper blade and slide it onto the arm until you hear it click into place, and then do the same routine with the second wiper blade. Lastly, once they're both replaced, go back inside your car and turn off Wiper Service Mode on your touchscreen, so the wiper blades return to their regular positions and you're ready to use them again.

How Do I... See How Many Miles My Tesla Has On It?

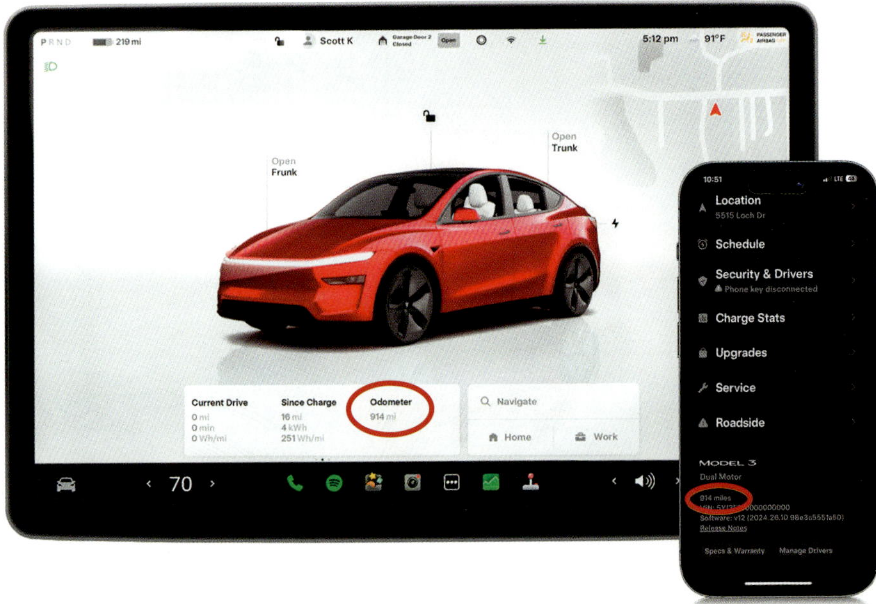

There are two places to find this one: The first is in the mini-panel (Tesla calls them shortcut "cards") at the bottom of the left side of your center touchscreen. You know the illustration of your car on the left side? Just swipe to the right and one of these "cards" (mini-panels) appears with your current trip info. But, if you swipe down through these cards, you'll come to one that has your odometer reading. You can also find this in the Tesla app on your phone. When you launch the app, on the main screen, scroll all the way to the bottom where you'll find your VIN (vehicle identification number) and your current odometer reading.

How Do I… Reboot My Tesla's Software?

If your Tesla starts acting squirrelly, the fix is usually the same as it is with your phone—reboot it. All you have to do is press-and-hold both the left and right scroll buttons on your steering wheel until your center touchscreen turns black and reboots from scratch. Just like on your phone, that'll usually do the trick. (*Note:* See page 190 for more on software updates.)

Chapter 12

How to Deal with the Haters

Owning a Tesla Does Not Come without Its Challenges

Okay, so we established in a previous chapter intro that my book publisher doesn't read these chapter intros (or any of the book, for that matter), or they would've clearly stepped in as the voice of reason, and just said, "No. No," and put a stop to the whole thing. But, they didn't, so I'm pretty free to talk about whatever we want (and by we, of course, I mean me, by which of course, I mean us, as in the colloquial us). Anyway, it appears our time together is almost over, so I want to share some parting wisdom. Now, I could lay some Tesla ownership wisdom on you, but the rest of the book is already all about your Tesla, so I think those might be "empty calories," if you know what I mean (I don't actually know what that means, but I heard it on an infomercial and you can't say it on TV if it's not true). Anyway, what I'm going to share in our brief, final time together is some stuff that can make a difference in dealing with the EV haters you're bound to encounter, and how to enhance your overall Tesla life. So, think of me as your "Tesla Life Coach." You can call me "Coach Kelby," or "Coach K" is good, too. Maybe even just "Coach," which has a nice ring to it, but yet, it's kind of impersonal, if you ask me. The important thing isn't what you call me (let's go with "Coach K"), it's how we spend this time together and the amount of wisdom I'm able to impart in the short time we have left, so I think we'd be wise to focus on that. Let's really zero in on what matters. Of course, I know there's so much more to life than just Teslas (great as they are, and granted, they are plenty great). So, I don't want to waste another moment on Tesla stuff when I could be sharing the single greatest lesson I've learned in life. If I had to point to one thing, one guiding light, one secret that made all the difference in my career, building a successful business, raising a wonderful family, and enjoying the peace that comes with an abiding faith while receiving total consciousness, I would have to say (and I don't share this lightly because this is a very personal message from me to you, but I say this without hesitation, and I'm holding absolutely nothing back) that would be....

How Do I... Reply to "You Can't Go Far in an EV"?

©VICTORIA WHITE

One of the first things people will tell you is that "they" can't drive an EV because it doesn't have enough range. This is not your problem, but they want to tell you that it is. The fact is, according to Kelley Blue Book, the average American drives 39.7 miles a day. Every Tesla ever built, has much more range than that. The lowest-range Tesla being sold today is the Tesla Model 3 rear-wheel drive, which has an EPA-estimated range of 272 miles. Then they'll tell you, "Well, you can't drive cross-country." The fact is people drive EVs cross-country every day. And, with a Tesla, it's even easier thanks to the Supercharger network. They'll also say things like, "Well, I wouldn't buy an EV unless it had 500+ miles of range and only cost $20,000." Don't argue with this person. They are just inventing scenarios to make it okay that they don't have an EV. If we believed the EV haters, we would believe that they drive 400–500 miles a day and drive cross-country every week.

How Do I... Reply to "EVs Are Just as Bad for the Environment as Gas Cars"?

GENERATED USING ADOBE INDESIGN AI

This is false and has been debunked over and over again. The premise for this is that the materials used to make the battery are sourced from third-world countries and these lithium mines are bad for the environment. There is some truth to this in that, yes, lithium mining is not great for the environment, but that's only part of the story. There is a one-time impact on the environment from this manufacturing process. However, the negative impact is usually negated within the first year of driving by the fact that an EV is not putting carbon emissions into the air. Let's also keep in mind that manufacturing an ICE (internal combustion engine) vehicle is far from a clean process—it's a dirty vehicle and gets dirtier as it ages during its lifetime.

 For more on this: **https://kel.by/earthjustice**

How Do I... Reply to "EVs Take Forever to Charge"?

©VICTORIA WHITE

It's true that you can fill up an ICE (internal combustion engine) vehicle in 5 minutes. It's true that charging an EV usually takes longer than 5 minutes. However, most EV drivers charge at home overnight or at work during the day. It's not like you're sitting there waiting for it to charge, outside of a long road trip, and even then the average charge takes 10–20 minutes at a Supercharger/Level 3 charger. Since I charge at home 99% of the time, I tell them that you spend way more time at gas stations than I do waiting for my Tesla to charge.

How Do I... Reply to "Replacing the Battery Costs More Than the Car"?

©VICTORIA WHITE

This is another misnomer that EV haters like to tout. It's true that a battery replacement can be expensive—replacing the main battery in your Tesla can cost between $5,000–$20,000. However, they are rated to last 10–20 years. There's also an 8-year-battery warranty on every new EV sold. I don't keep my cars for 8 years, so I actually never think of this. However, if I had a 10–20-year-old Tesla and I could spend $5,000–$20,000 for a new battery to get another 10–20 years with the savings in gasoline over the years, I'd call that a win.

How Do I… Reply to "What If the Grid Goes Down?"

©ADOBE STOCK/LEO

If there is a power outage due to a problem with the grid, then, in most cases, you can't get gas either since the pumps at the gas stations are powered by electricity. Also, most EVs have a high rate of charge each day from charging up the night before. This means that if the power went out on any given day, you'd probably have enough charge to last a few days or you could drive to a location that has power and charge up. Lastly, the grid is not the only way to get electricity. Some Superchargers have solar canopies and others have large on-site batteries. I have solar at home. A temporary grid outage isn't a real concern.

How Do I... Reply to "They're Dangerous Because They Catch on Fire All the Time!"?

©VICTORIA WHITE

Every time an EV catches fire it makes the headlines. I have a theory as to why this is the case: The reason an EV fire is news is because it doesn't happen that often and the media wants to grab attention, clicks, and views. We pass ICE (internal combustion engine) vehicle fires all the time and as long as no one was in the car at the time, we don't give it a second thought. So when EV haters hear about yet another EV fire, in their minds, it must happen all the time. People have been told that they are risking their families by charging their EVs in their garages. Now let's talk about the actual numbers: As of March 2024, Edmunds reported that electric car fires have averaged around 20 per year over the past three years, out of about 611,000 electric vehicles. In comparison, gasoline-powered cars, which number about 4.4 million, have had around 3,400 fires during the same time—20 versus 3,400. I'm pretty sure that the vast majority of those 3,400 ICE vehicle fires were not deemed newsworthy, but I bet that all 20 of the EV fires were all over the news. (By the way, this is not a real Tesla fire, but if you listen to the haters this is what happens all day everyday.)

How Do I... Reply to "You Can't Drive an EV in the Winter/Cold"?

©VICTORIA WHITE

Last year, there was a story about a couple of Supercharger locations in Chicago that had some chargers go offline and people couldn't charge. There were also a few drivers that didn't precondition their batteries (see page 106 for more about this) and when they arrived at the Superchargers in the frigid temperatures, they wouldn't charge. Yep, these things happened that one day in that one city. It's true that cold temperatures have a negative impact on EVs—you could lose as much as 30%–40% of your range in very cold temperatures. This has been a problem with batteries since day one, including the batteries in ICE (internal combustion engine) vehicles. Having lived in Michigan, I've woken up many a day where my ICE vehicle wouldn't start and required a jump. Here's the reality: In Norway, over 80% of the new vehicles sold are EVs. In Norway, the average temperature in the winter is –6.8 degrees Celsius (19.76 degrees Fahrenheit). With these cold temperatures, you would think that Norwegians would stick to ICE vehicles, but nope, they prefer EVs. Driving an EV in the winter works just fine, but does require some driver education (again, see page 106).

How Do I... Reply to "The Electricity for Your EV Comes from Coal"?

You'll hear that your EV isn't really helping the environment because the electricity comes from dirty coal anyway. The fact is that power companies have been moving to cleaner sources for years. On average, only about 20% of the country's electricity comes from coal these days. Even when it does, EVs are still cleaner than their ICE (internal combustion engine) vehicle counterparts. Also remember that, unlike gasoline, electricity can come from several cleaner sources like solar, wind, and hydroelectric. Lots of EV owners also have solar at home and Tesla has solar-powered Superchargers. There is no clean way to produce gasoline.

Breakdown by state of how electricity is generated: **https://kel.by/afdc_energy_data**

For more on this: **https://kel.by/forbes_electric_cars**

How Do I... Reply to "Good Luck Taking a Road Trip"?

©VICTORIA WHITT

People hear "200–300 miles of range," "24 hours to charge," and "not enough EV chargers" and automatically assume that road trips are difficult at best and impossible at worst. Having personally driven my Tesla from Atlanta to Tampa (456 miles) and back, as well as from Atlanta to Cincinnati (461 miles), I can report firsthand that these trips were a breeze and about as far as I would drive any car. I put my destination in the Tesla navigation system, get on the highway, and engage Full Self-Driving (Supervised). These have been very easy trips. I had one to two short Supercharger stops and even if I didn't have to stop to charge, I would have stopped anyway to use the restroom and grab a bite to eat. These trips were 100% problem-free. I have friends who have made multiple Tesla and other EV trips from Georgia to California (2,481 miles) and they did it with no problem. With a Tesla, a road trip is easy. With other brand EVs that can't yet use the Supercharger network, it may be a bit more challenging, but certainly not impossible (the other networks don't have as many Level 3 chargers and not as many stalls at each location).

How Do I... Reply to "Your Electric Bill Will Skyrocket"?

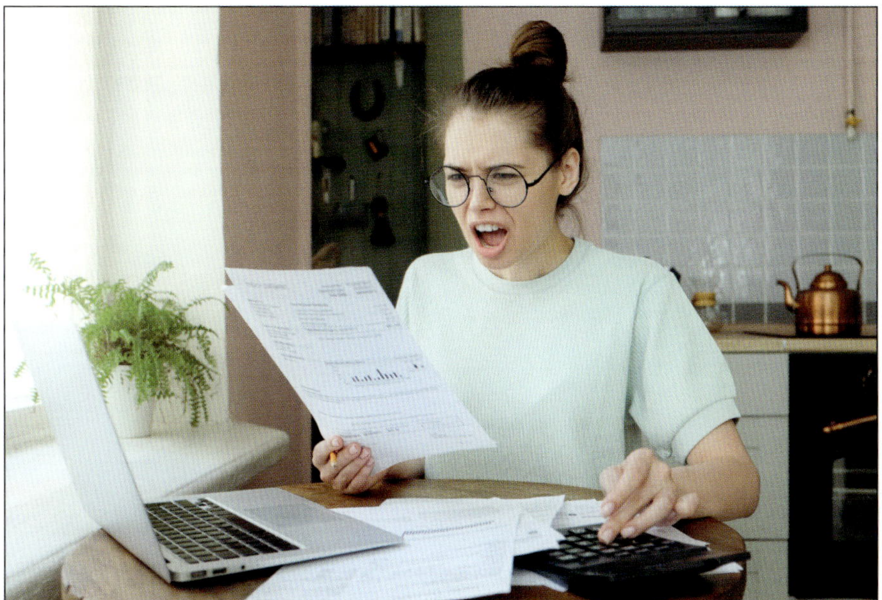

©ADOBE STOCK/DAMIR KHABIROV

If you charge at home, then you'll likely use more electricity than you were before you got your Tesla. However, outside of California (where the electricity rates are more expensive than any other state), the increase in your electric bill will pale in comparison to what you were paying for gas. If you're ever in doubt, you can put your home electric rates in the Tesla mobile app and it will estimate how much you've paid in electricity at home, work, and public chargers. It will also show you an estimate of how much you've saved in gas (see page 178 for more on this).

How Do I... Reply to "There Aren't Enough Public Chargers"?

©ADOBE STOCK

There are gas stations everywhere—sometimes two or three at the same intersection or right next to each other. This is because drivers of ICE (internal combustion engine) vehicles have...(wait for it) range anxiety! That's why there are so many gas stations. They never want to not be able to fill up. Therefore, when they see far fewer public chargers, they will tell you things like, "Good luck finding a charger!" While we could always use more fast public chargers, there do not need to be as many public chargers as there are gas stations. There's a big reason for this and it's something that the haters never talk about: it's the fact that the vast majority of EV drivers charge at home/work. Ask them, "If you had a gas pump in your garage, how often would you stop at a public gas station?"I only need a public charger on long road trips, and for those, I have plenty of Superchargers along my route.

How Do I... Reply to "The Grid Can't Handle Everyone Getting an EV"?

©ADOBE STOCK/@DESY

There are times during the summer months that some cities struggle with keeping up with the demands on the electric grid. That's a fact. However, power grids are always being upgraded to support the increased demand. The stress on the grid comes more from everyone running their air conditioners all day, than from EVs charging. Also, most home EV charging happens at night when the demand on the grid is much less. In fact, experts see EV batteries as part of the solution. They help to reduce planet-warming emissions and can add needed flexibility to electric utilities that are sure to come under more strain as global temperatures continue to rise. The world isn't going to suddenly switch to 100% EVs overnight, but as more EVs are being sold over time, the grid is being upgraded as well.

How Do I... Reply to "If You're in a Traffic Jam, You'll Run Out of Juice"?

©ADOBE STOCK/MIKE DOT

When your EV idles, it uses very little battery charge. In fact, people have been in traffic jams for hours in their EVs and hardly had any drain—yes, even with the heat/air conditioning on. When your EV is moving, that's when it uses more battery charge. Teslas are very efficient. This is why you don't see a bunch of news stories with headlines like "Several EV drivers stranded in traffic on the I-5" every day.

How Do I... Reply to "Nobody Wants or Can Afford Them"?

©ADOBE STOCK/CAROLINA JARAMILLO

There was a recent decline in EV sales and, of course, this must mean that, "See, nobody wants them!" At the same time, the Tesla Model Y was the number one selling car in the world in 2023. Not the number one selling EV, but the number one selling car period. Clearly, people want EVs, but the decline is likely due to other manufacturers overestimating the demand, and then pulling back when they didn't meet their numbers. In terms of the "no one can afford them" remark, well that is from a misconception that all Teslas are $70,000–$100,000+. The first Tesla was a Roadster ($80,000–$100,000), and the second one was the Model S ($57,400). These were never meant to be mass market EVs. Tesla always planned to introduce a more affordable Tesla, and that became the Tesla Model 3. A brand-new, base Tesla Model 3 is $38,990. That's not including tax incentives or gas savings. The average price of a new car in the US is $48,008. So, while they will argue that there are cheaper ICE (internal combustion engine) vehicles, you could argue that there are cheaper EVs, too. While EVs can cost more to buy up-front than ICE vehicles, the cost of ownership is usually much less over the life of the vehicle.

Index

J

K

L

M

N

O

P

R

S

T